Technological Utopianism and the Idea of Justice

"As in his previous work, Sand manages to redraw the boundaries of the bourgeoning debate about technological futures by focusing sensibly on justice issues in this field. He asks us to interpret technological futures as utopias and considers their potential to improve our understanding of justice's scope and location. Another highly original approach!"

—Prof Armin Grunwald, *Ethics and Philosophy of Technology at the Karlsruhe Institute of Technology (KIT) and Head of the Institute for Technology Assessment and Systems Analysis (ITAS)*

"Sand makes an intriguing proposal – can technological utopias teach us anything about the scope and location of justice? While many technological visionaries will hope so, certainly no one has ever so ambitiously defended that claim before him…"

—Prof. Dr. Vincent Blok, *Professor in Philosophy of Technology and Responsible Innovation (Wageningen University), Professor in Philosophy of Data Science and AI (Erasmus University Rotterdam)*

Martin Sand

Technological Utopianism and the Idea of Justice

palgrave
macmillan

Martin Sand
Values, Technology and Innovation
Delft University of Technology
Delft, Zuid-Holland
The Netherlands

ISBN 978-3-031-75944-4 ISBN 978-3-031-75945-1 (eBook)
https://doi.org/10.1007/978-3-031-75945-1

The open access fees of this publication have been generously covered by the TU Delft Open Access Fund (Prometheusplein 1, 2628 ZC Delft, The Netherlands) and the Department of Technology, Policy and Management of TU Delft.

This Palgrave Macmillan imprint is published by the registered company Springer Nature Switzerland AG
The registered company address is: Gewerbestrasse 11, 6330 Cham, Switzerland

For Jonas

Contents

1	**Why Utopia Instead of What Utopia**	**1**
	1.1 Transhumanism	16
	1.2 Virtual Reality	19
	1.3 Computerized Communication and Cyberspace	21
	References	25
2	**Perfectionism, Stagnation and Transcendental Theorizing**	**29**
	2.1 The Failures of Predicting and Remodeling Society	32
	2.2 Varieties of Practical Relevance	42
	2.3 Lock-In, Boredom and Stagnation by Boredom and Stagnation	53
	2.4 Untenable Epistemological Absolutism	59
	References	65
3	**Technological Anti-anti-utopianism**	**69**
	3.1 Technological Utopians Are Ruthless and Elitist	72
	3.2 Engagement with Unrealistic Scenarios Deflects from More Important Present Concerns	79
	3.3 Technological Utopias Fetishize the Technological and Are Socially Oblivious	86
	3.4 The Social Obliviousness of Technological Utopias Amounts to an Acceptance of Liberal Capitalism	93
	References	102

4 How to and Where to Justice 109
References 128

5 Justice in Technological Utopia 131
5.1 Immutable Human Biology and Its Impact on Talent and Equality 132
5.2 Material Scarcity—Or How Flourishing Can a Society Become? 137
5.3 Reconciling Pluralities of the Good Life in Virtual Reality 142
References 144

6 Conclusions 147
References 152

Index 153

List of Figures

Fig. 2.1	Various functions of utopianism	48
Fig. 3.1	Reciprocal relation between technology and socio-political institutions	93
Fig. 4.1	Reciprocal relation between individual behavior and socio-political institutions	115
Fig. 4.2	Reciprocal relation between individual behavior and technology	116
Fig. 4.3	Technology, social institutions/politics and individual behavior as constituents of justice	117
Fig. 4.4	Scope of justice	122

CHAPTER 1

Why Utopia Instead of What Utopia

Abstract This introductory chapter paves the way for a reconciliation of Utopian Studies, philosophy of technology and political philosophy. It introduces the general idea of vindicating technological utopias and engagement with them. The value question is being moved to the fore of the present essay, sidestepping the quest for a definition to unresolved methodological problems. A commonsensical idea of utopia is not a priori flawed; its value still remains unclear. Some paradigmatic technological utopias are introduced.

Keywords Defining utopia · Technological utopianism · Worthwhile engagement · Transhumanism · Virtual reality · Cyberspace

Technological utopias are both increasingly abundant and controversially discussed. The present essay aspires to contribute to these topical discussions. It will propose a modest vindication of technological utopianism suggesting that engagement with technological utopias is worthwhile, because they can advance our understanding of the scope and location of justice and, thereby, the idea of justice more generally. When considering justice and establishing theories of justice—what it is and what its location and scope is—theorists entertain background assumptions about aspects of human nature, ways of living and how the world is composed, which they consider as immutable and that underlie and pervade their

M. Sand, *Technological Utopianism and the Idea of Justice*,
https://doi.org/10.1007/978-3-031-75945-1_1

theories. Some such background assumptions that are often rather implicitly than explicitly accepted in theories of justice are some level of scarcity, human mortality, limited altruism, limited knowledge and information in decision-making processes and reasonable disagreement between political actors. Referring to such obstacles, to which a theory of justice would have to provide a viable response, Jacob Levy suggests: "Political organization and justice are about moral friction in the first instance" (Levy, 2016, p. 312). The aforementioned background assumptions cause such "moral frictions": How shall we distribute goods that are limited, is one of the most pressing questions that arises, if scarcity is taken as a background assumption. How can we peacefully move on with political and social life given the many reasonable, but incompatible answers to socio-political questions, is another one that naturally follows.

Even if we agreed that a theory of justice in an ideal world devoid of *all* of these constraints might be pointless, the question that arises is: Why shall we accept *those* background assumptions? How incompatible are political views and diverging views about the good life? Is human nature—*if* it were true that it is to only a limited degree altruistic—really immutable? Must scarcity and mortality be accepted as the backdrop for any theory of justice to commence? How insurmountable are those background assumptions really? Many technological utopias provoke us to question whether one or all of these background assumptions are indeed immutable and how our concept of justice would change, if they were not. Such provocation is illuminating and has positive value (and in some sense also practical value, as we shall see). Technological utopias estrange us from the present, which fulfills an important dispositional and cognitive function. They alter our perspective on readily accepted views of the world and the alleged immutability of many aspects of social and political life. This constitutes, in my view, the positive value of technological utopianism and provides a reason for scholars to engage with them.

Let us call this thesis in the following "WE" (worthwhile engagement). As we will see, such a view faces criticism—old and new—which shall be extensively discussed in this essay. Aside from traditional charges that utopianism is problematically perfectionist, recent criticism addresses contemporary technological utopias more directly. Technological utopias are supposedly driven not by subversive motivation—a desire to criticize and, thereby, illuminate the flaws of the reigning socio-political order—but by corporate interests. Rather than speculating about better ways of living, they are falling trap to a naïve technological determinism and materialism. They are allegedly socially oblivious and, therefore, in a particular way naïve or unrealistic.

As a response to these charges, I will explain how these narratives envision new socio-technical constellations in which justice emerges. They exceed the narrow dichotomy between individualist vs. collectivist utopias, placing technology as another possible constituent towards an ideal society. Oftentimes they unravel creatively how particular values might be promoted or undermined through the interaction of technologies, individual behavior and socio-technical institutions. Because of this, they illuminate, can provide orientation and incite or deter action. (Although it is by no means straightforward, which action they should incite. We will later consider, how approximating the realization of utopia can cause overall decrease of flourishing or well-being for society.) This does not mean that such technological utopias or utopias in general should be the *only* object to study for scholars interested in justice and politics. It is not contradictory to claim equally that to make more practical socio-political decisions on technological change, perhaps more realistic views are at least equally or even more important.

In sketching this view of WE, I say purposefully that technological utopias *can* advance our understanding of the location and scope of justice and that they *oftentimes* do—acknowledging that not all of them do this. This feature of my view is not a bug. The fact the reading some books is a gigantic waste of time does not undermine that reading books is by and large a valuable thing to do. So, my defense rests not so much on the *form of technological utopia* but on their possible *content*. The content of some utopias constitutes their cognitive and dispositional power. Hence, this vindication of technological utopianism is modest in that I do not aspire to salvage all existing technological utopias. There are many that are badly motivated, one-dimensional and uninspiring. These are not worthy of engagement. I will also not defend these narratives *as such*, but in their capacity *as utopias*. Hence, we must not affirm associated prophecies, nor seek out for direct answers on how to reshape or govern our present societies—though, we might excavate here and there some interesting ideas about that, too. Rather, we should treat them as visions that prompt important questions and crucial reflections on how to shape the world to make it more just and livable.

The essay's main thesis (WE) will be developed in the following steps: Rather than providing a definition of technological utopianism, I will commence by outlining some problems of defining abstract concepts such as "utopianism" in the social sciences. A too narrowly constructed definition comes with the heavy theoretical cost of defining the visions we are interested in, and which are almost unanimously considered as utopias, out of the purview of this discussion. Without endorsing her views, I will articulate my sympathies for Ruth Levitas' account of utopias as expressions of the desire for a better way of living and being (Levitas, 2011), as it avoids the aforementioned problem. I will then introduce some exemplary technological utopias that are prominently debated nowadays. In Chapter 2, by discussing some traditional anti-utopian arguments, we will make inroads toward a more positive picture of utopianism and interrogate what they are good for in general. Chapter 3 will expand on this anti-anti-utopian strategy and discuss in more detail criticism that is more specifically directed at technological utopias (their social obliviousness, the ruthlessness of their promoters and their woeful alliance with prophecy). In Chapter 4, I introduce the idea that justice has a location and a scope and how technological utopias challenge the traditional dichotomy between individual behavior and socio-political institutions as primary locations for justice. In this way, the present essay delineates how technological utopias fulfill the cognitive and dispositional function that makes utopias in general valuable vis-à-vis our understanding of justice. This general idea will be spelled out with some examples in Chapter 5. In my final conclusions in Chapter 6, I will renounce two more objections to my account and allude to some possible future avenues for research.

So, what is utopianism? And in particular: What is technological utopianism? Establishing a concise definition of utopianism is strenuous, as other authors have pointed out before (Thaler, 2018). A key problem to start with is that the fields that are interested in technological utopias (Utopian Studies, philosophy of technology and political philosophy) have no unanimously established method to arrive at definitions. Any such method has caveats. In philosophy, when it comes to definitions of abstract concepts and questions such as "Am I my brain?" or "What is agential integrity?", thought-experiments and examples are conjured and applied to test our intuitions and folk concepts: If someone knew everything about my brain, would she know everything I know, everything *about* me? Would it threaten my integrity, if one were to force me to kill someone, to save five others from being killed? However, cases such as these have caused an intractable literature on the reliability of intuitions, wondering whether intuitions are ever coherent and how they can be affected by experimental framing (Lillehammer, 2011). Hence, there is doubt whether such exercises of triggering intuitions can be reliable guides to validating abstract definitions and theories. What other ways do we have in the social sciences and humanities to establish definitions?

Another possibility to reach a definition is to consider a concepts' etymological roots and identify original, paradigmatic or canonical usages or cases to analyze and filter a concepts' main ingredients. Krishan Kumar has gone a long way to suggest—strongly based on a reading of Thomas More's *Utopia* (Kumar, 1991)—that "Utopia is nowhere (*outopia*) and it is also somewhere good (*eutopia*). To live in a world that cannot be but where one fervently wishes to be: that is the literal essence of utopia" (p. 1). He continues to suggest that "[...] Utopia describes a state of impossible perfection which nevertheless is in some genuine sense not beyond the reach of humanity" (p. 3). He furthermore adds the following functional dimension:

> This [criticism of society] could take the form of an argument from general principle. But More rejects this in favour of theory 'by demonstration'. This is the novelty of Utopia - and of utopia as a form of social theory or 'theoretical practice'. More does not merely assert, he shows the systematic interconnectedness of evils and their remedies by portraying a fully realised alternative social order. [...] It is in this new perspective that utopia is most truly subversive. (p. 3)

Kumar establishes here a definition that alludes to three dimensions; utopias content (description of a perfect state), its form (a state that is one of impossible perfection, but not entirely beyond reach) and its function (the criticism of the reigning social order as a "theoretical practice"—its subversiveness).[1] Given the focus of this essay, one

[1] Lucy Sargisson suggests, in contrast to Kumar's reading, that perfectionism is not in the fore of utopias but of *eutopias*. Furthermore, according to her, many contemporary utopias "disrupted the boundary between eutopia and dystopia; placing both eutopic *and* dystopic visions inside the same text" (2012, p. 10). Reflecting the ambivalence and complexity of utopian world making within utopias themselves, is for her—following Frederic Jameson (2005, p. 217)—a core element of contemporary utopianism, exemplified in the works of LeGuin and Kim Stanley Robinson. For her, the element of subversiveness, the fact that "utopianism stems from discontent with the now and utopias always contain criticisms of their present" (ibid.) is a necessary element of any definition of utopianism. It is noteworthy, in light of my following argumentation, that her definition is established without commitment to one of the two roughly introduced methodological doctrines. She

would merely have to add that such state of impossible perfection ought to be brought about primarily through technology rather than societal reform to approximate a definition of technological utopia. Hence, perhaps, in a technological utopia, the subversive aspect would not have to be very pronounced, given that many technological utopians care little about the socio-political context in which they promote their "state of perfection." An analysis of this interrelation is, of course, what we will elaborate in later stages of this essay. Such rough definition might indeed satisfy most scholars of technological utopianism and could, therefore, be pragmatically adopted.[2]

Kumar's definition is obviously established via a close reading of Thomas More, although he does not stick very strictly to it in his subsequent outline.[3] But, why choosing Thomas More as a point of reference

contests conceptions of utopia as comprehensive or "totalizing" with selected examples such as Kim Stanley Robinson *Mars Trilogy*. No further defense, however, is provided for this commitment and naturally one wonders: Why should we take Robinson's *Mars Trilogy* as a utopia in the first place, if other definitions would rule it out?

[2] Howard P. Segal, for instance, devices a notion of "genuine utopias" as containing some sort of "perfection [entailing] a radical improvement of social conditions as compared with non-utopia [...]" (2006, p. 2). Additionally, he suggests that utopias ought to be comprehensive to a certain degree: "genuine utopias seek change everywhere, in the basic structure of society" (ibid.). His history of *Technology and Utopianism*, however, is *not* the history of *technological utopianism*. He suggests many utopians in whose work technology plays a central role, in particular early utopians such as the Pansophists (Campanella, Andreæ and Bacon), in which, however, "technology is only a means to an end and not, as in full-fledged technological utopianism, an end in itself" (p. 20). Hence, he introduces here a distinction between utopias, where technology plays a heavy role in bringing about diverse societal goods and those in which "technology becomes and end in itself", which seem to denote technological utopias proper. Technology as an end in itself, however, seems an excessive characterization for any existing narrative: What would such narrative look like? If we cannot refer to any, we might very well have invoked a strawman. We will discuss the alleged social obliviousness of some current technological utopias in more detail in Chapter 3. It is in any case remarkable that here—as in many other works—the question "What is utopia?", is at least tentatively answered, while the question "What is *technological* utopia?", almost never is.

[3] It is widely held that—*contra* Kumar—More's utopia is full of ambiguities and irony (Bruce, 2008; Sargisson, 2012, p. 14). When discussing utopian practices and communities and their oftentimes inexistent or only implicit relation to utopian literature, Kumar suggests that "these examples [of utopian practices] suggest something of the limits as well as the possibility of establishing some determinate connection between utopian thought and utopian practice. [...] Utopias are not written to be realized, or at any rate in any direct literal sense. Their ideal of perfection is theoretical; their writers may be quite indifferent to the problems of achieving their goal in practice [...]" (1991, p. 72). Thus,

and not, for instance, Plato's *Republic*, as other authors did (Sargisson, 2012, p. 7)? And, how can we capture those utopian practices that do not have literary foundations—that are not theoretical practice, but actual practice such as the communities of Drop City in Colorado, Twin Oaks in Virginia or the Whole Earth community with and without expansive theoretical backdrops (Sargent, 2010, p. 41; Turner, 2006)? These practices seem to be part of *The three faces of utopianism*, as Lyman Sargent has prominently outlined (Sargent, 1994).

These questions underscore the shortcomings of an attempt of defining via etymology: Naturally, such route can be challenged by recent examples of utopianism that indicate the mutability of the concept and the genre. To make an analogy: If love nowadays is practiced, felt and understood very differently from how it has been felt, practiced and understood in the Middle Ages, why should early medieval literature's conceptualizations of love (if we were to take those, for the sake of argument, for the concepts' etymological origins) guide our current definition of love? In general, the mutability and dynamism of abstract concepts over longer time scales poses threats to any definitional activity whose intent is the creation of a static, stationary target. It is a particularly vulnerable undertaking if it shall rest on what are allegedly foundational texts.

That this is a particular challenge for a dynamic, changing and evolving tradition, whose emanations are shaped by mutable socio-political contexts, has been clearly identified by Ruth Levitas. She warned of definitions that "obscure the variations in the utopian genre" (Levitas, 2011, p. 8). This obscuring is nowadays seen as particularly problematic as utopian scholarship has become self-aware of its earlier Eurocentrism, inhibiting severe blind spots on utopian traditions elsewhere, blind spots which have been alleviated only in recent years (Longxi, 2002; Sargent, 2010; Segal, 2012). Levitas' widely hailed attempt toward establishing a definition via a comprehensive survey of existing accounts is guided by the conviction that a definition must distill "the substance of a concept" ("something which remains constant"), while remaining flexible enough to encompass its various manifestations—practices and forms that are generally considered manifestations of utopianism. Her account is motivated by the fear that "without a definition, it is difficult to establish exactly what we are looking at" and that "without agreement as to what

instead of expanding his definition, he is clearly prepared to bite the bullet here and exclude utopian practices from the concept of utopianism.

the proper limits [of the concept], if any, are, there is danger of researchers in the area making arbitrary and subjective selections of material; or, even if they are clear and methodological in their own use of the term, using it in an idiosyncratic way and talking past each other; or, indeed, wasting too much time arguing over what is or is not utopian" (p. 4). Constructively, she later argues—and it is worth citing this longer passage in its entirety—that:

> [...] all definitions in terms of form, function or content are problematic. Not only do they place limits upon what may properly be regarded as utopian and thus upon the field of enquiry itself, they also obscure variations in the utopian genre. In order to make such a claim, one must of course be able to locate something which remains constant while content, form and function vary. This element, I would argue, is that of desire – **desire for a better way of being and living.** [...] where such desire is expressed – and the scope for this will itself be historically variable – it will not only vary markedly in content but may be expressed in a variety of forms, and may perform a variety of functions including compensation, criticism and the catalysing of change. The most useful kind of concept of utopia would be one which allowed us to explore these differences and which ultimately might allow us to relate the variations in form, function and content to the conditions of the generating society. At the same time, it would not exclude from the field of utopian studies any of the wide variety of related work that currently is defined by practitioners as part of the field. (Levitas, 2011, pp. 8–9)

We see immediately how this approach contrasts with the paradigmatic-etymological one of Kumar: Kumar's is a top-down approach, starting with a paradigm case of utopianism to then see which particular instantiations fall under that conceptual umbrella. The latter—Levitas' approach—is a bottom-up approach; it presumes the mapping of the field and the distilling of one or more features that all those surveyed narratives have in common. Levitas claims to have found this common denominator in all utopias in "the desire for a better way of being and living". I have already outlined the problem of the top-down approach above. What about the bottom-up approach? Can we embrace it as way forward to establish a definition of utopianism? Without doubt, the definition of utopia as "desire for a better way of living or being" has been highly influential and seems strongly appealing, as it comprises a plethora of different political tractates, literary documents, but also experiments in communal living and other practical endeavors. Without committing myself to this idea as the definitive account of utopianism, I will regularly refer to it in the following when we can detect its presence in expressions and passages in some technological utopias.

Again, in line with the purpose of the present essay, one might be tempted to take this account and endorse it based on the scholarly authority that Levitas' account enjoys among utopian scholars and merely add that technological utopia denominates those desires for whose fulfillment technology is seen as a primary means. Hence, one would merely add a minor form-requirement to Levitas' broader definition of utopianism and would end up with a decent definition of technological utopianism: Technological utopianism could thus be understood as a "desire for a better way of living or being which are to be achieved primarily with technological means". This definition might satisfy most of my readers.

The ingenuity of Levitas' analysis notwithstanding, however, there are at least two problems that challenge this bottom-up approach. First, while it is fear of arbitrariness and imprecision (see above) that motivates Levitas' efforts toward a definition, it is by no means clear that her definition has resolved that problem or that, in fact, any definition can resolve this problem. Will it help the utopian scholar to decide whether the movie *Interstellar* presents a utopia, if utopia is understood as a "desire for a better way of living and being"? It is quite obvious that a definition of utopia as a "desire for a better way of living and being" stemming from an interest in clarity and demarcation merely generates a need to define the concepts "desire," "living" and "being," whereby the latter poses problems of a magnitude that might even eclipse those of defining utopia itself.

Arguably there are many (arbitrary) definitions that would solve the arbitrariness problem that motivates Levitas' account much better (e.g. that utopias always display perfect states, which *Interstellar* obviously does not).[4] This reveals an underlying antagonism regarding the various goals in this account and opens the broader question of the purpose of defining utopianism: What purpose ought such definition fulfill and are the goals of finding a bottom-up, encompassing definition not detrimental to the goal

[4] We see that even with such slightly less ambiguous and straightforward definition, a debate would swiftly open about what one can consider "perfect" here. A retreat battle has begun.

of finding a definition that helps scholars to distinguish sharply between materials that fall inside or outside that category? Levitas' given answer seems in this regard inconsistent and unjustified.

Second, the choice to find a definition that "would not exclude [...] any of the wide variety of related work that currently is defined by practitioners as part of the field" might very well beg the question of how a definition of utopianism ought to proceed. Why should we include those manifestations in our forming of a definition? Should we not rather exclude them because paradigm cases suggest that they do not fall within the scope of the concept? Such question begging is perhaps not ultimately unjustifiable, but *prima facie* problematic.

Whether either schematically introduced bottom-up approach or the top-down approach is preferable, is an intractable question and it is unlikely that we will reach an acceptable agreement concerning this question—certainly not within the scope of the present essay. Furthermore, aside from these two schematically sketched approaches there are others, to which one would have to do justice, and this too is unfortunately beyond the scope of the present project.

My own view, in any case, is that a definition of technological utopianism is also not necessary. For the purpose of the present essay, the definitory question "What are technological utopias?" is less interesting than "Whatever they are, what—if anything—makes them worthwhile of engagement?" To stick to the previously introduced analogy, we do not need to establish a definition of "book" to advance the general claim that reading is by and large something good. We would need it only perhaps, if we wanted to argue that reading this or that piece of literature is worth reading, because it is a book—which is an odd argument indeed.

Hence, the main focus here is the normative question, what value technological utopias have—what is good (or bad) about them? This question might be to some extent tied to questions of definition, yet it is still largely distinct from them. For instance, if utopia were, as common parlance suggests, *unrealistic dreaming*, then one could immediately—if not outrightly—reject them, or at least concede that they hold a massive flaw, namely, that it is impossible to realize them. If utopias were *by definition* unrealistic, they might still entice some people, provide hope or stimulate thought, but their import for concrete political action would be markedly diminished. In fact, one might think that the radiating spell of utopianism that fuels some peoples' hopes in a better future, is dangerous insofar as it must lead to disappointment, when those people find out that their hopes were unrealistic from the get-go. Disappointment is a possible but not a necessary result of utopian imagination. Even then, however, one will have to concede that—despite the risk of disappointment—the unrealistic, imaginary reconstruction could be valuable in that dissociation from the burdens of the present can offer temporary relief from anguish and agony and that taking steps toward realizing an unrealistic future, might well lead to concrete betterments in the present.

Or, to provide another example, if utopia were—as Howard Segal claims with reference to the work of George Kateb (Segal, 2012)[5]—the envisioning of a comprehensive remodeling of society, the fear of centralized, comprehensive state intervention into everyday life seems not unfounded. Karl Popper summons the fear of totalitarianism as closely tied to the very idea utopianism suggesting that "[h]olistic or Utopian social engineering, as opposed to piecemeal social engineering, is never of a 'private' but always of a 'public' character. It aims at remodelling the 'whole of society' in accordance with a definite plan or blueprint [...]" (Popper, 2012, pp. 61 f.). Of course, Popper here dangerously equates the remodeling of the entire society (breadth), with a remodeling of all of society (depth) and claims later, confusingly, that this is not the crucial difference between utopianism and piecemeal engineering. The crucial difference according to Popper is to be found in how cautious such remodeling shall proceed. As piecemeal social engineering allegedly proceeds in incremental and in minor steps, always considering reversibility of interventions, it would be the preferable approach—one that is less prone to the risk of causing collateral damage. We will return to this later on and see that, luckily, it is neither obvious that being concerned with the remodeling of the "whole society" is problematic, nor that being practical in a direct political sense is indeed a necessary feature of utopianism. Though, we will also see in Chapter 3 that many technological utopians pursue the realization of their imaginaries as practical endeavors and such practices, I shall concede, require careful scrutiny.

Therefore, it is not obvious following any ordinary, folk concept of utopianisms that it is flawed and valueless. Oftentimes, in common parlance, utopianism is understood as demanding the impossible and the

[5] Segal writes: "[...] not only their precise contents but also their comprehensiveness further characterize genuine utopias, which seek changes in most, if not all, areas of society. By contrast, false utopias seek changes in only one or two components, such as schools, prisons, diet, or dress" (2006, p. 6).

pursuit of unrealistic goals. To demand the pursuit of something that is literally impossible to achieve, would be irrational indeed. Ought implies can, after all. Though, it is unlikely that utopianism is by definition irrational. Two different resolutions are possible—and both have some appeal to them: In recourse, we could suggest that utopias either do not *demand* the impossible or that they *do not portray* it. These two aspects do not need be part of utopianism conjunctively. Constraining the definition to either one of those readings would be sufficient to reject the claim that utopias are per se irrational. Most utopian scholars side with the former reading: They suggest that utopians do not actually *demand* the impossible, or in any case do not necessarily need to demand the impossible, to achieve something that is otherwise valuable, namely imaginatively work out alternative ways of living and being that fulfill various functions in societal and political debates. This is also clear for Ruth Levitas, who writes that "it is possible for an imagined world to carry out any of the functions of compensation, criticism or change without being possible [...]. To function as criticism or compensation, utopia does not even have to be believed to be possible" (Levitas, 2011, p. 219). Aside from the relevant and valuable functions that even unrealistic utopias can fulfill (e.g. Land of Cockayne), one must in general remain reluctant to maintain the rather contentious distinction between realistic or possible and unrealistic or impossible proposals for better ways of living in this context. Too many allegedly realistic proposals have failed due to unforeseen realities and too many allegedly impossible reforms (abolishment of slavery), were realized although they were deemed impossible by many for decades and centuries until then (Estlund, 2020, p. 259). The costs, therefore, of rejecting utopianism based on an alleged display of something impossible—being it a way of living, a technology or a political ideal—underestimates the causal role that the conjuring and displaying of those very images have in making them possible. In other words, conjuring what seems impossible at one point in time might very well emerge as the *sine qua non* for achieving it in at a later point. Because of this causal role of the imagination in actualizing possibilities that have previously been unthought of or deemed impossible, utopianism is seen by Paul Tillich as always being suspended between possibility and impossibility (Kumar, 1991; Sargent, 2010, p. 91).

To repeat, there is no reason to think that any definition of utopia that liaises closely with our folk concept of it, unravels an immediate paradox or helpless irrationality. No matter how utopias are defined—whether as experiments in living better, as education of desire or as formulations of ideal principles of justice—whether utopianism has any value and is worth of intellectual engagement, is a question that is not a priori settled. It is this question about value that motivates the present essay.

Surely, the pleasure or joy that some authors and thinkers experience establishing those schemes seems a *pro tanto* value to permit them to do so, and to some the question is not one that needs to be posed: Still, does that give scholars a reason to investigate their products and to consider what they propose, to study them, to analyze their feasibility and desirability? That is not obvious. To restate: The view brought forward here is that technological utopias are valuable and can be worthy of engagement insofar as they can illuminate our idea of the scope and location of justice, and some technological utopias do this.

Though, we have glanced over some aspects—perhaps even prominently associated with the folk concept of utopianism—that are worrisome for anyone with the impetus to defend utopianism. We should pause a moment to reconsider those. Intriguingly, Lucy Sargisson suggests that perfectionism should not feature in a *definition* of utopianism, not only because it fails to capture the essence of many contemporary utopias that are ambivalent (in line with Levinas' request to do justice to utopian variability), but also because perfectionism is in various ways morally and politically problematic. She writes:

> [...] it is my contention that perfection is not actually a defining feature of utopianism. My claim stems partly from scholarship: many canonical utopias (including More's *Utopia*) contain imperfections. And it is partly normative: utopias *should not* seek to create a perfect world. [...] The consequences [of perfectionism] are intellectually and politically lethal. (Sargisson, 2012, p. 14)

Sargisson seems to suggest here that a normative demand—the value of imperfection—shall have an impact on utopias' definition. My argument does not entail this: Those technological utopias that do *not* advance our understanding of the scope and location of justice—which I purport to be a valuable feature of many technological utopias—do not thereby disqualify as technological utopias. Assuming this would be odd: Why should a definition of "chair" try to exclude electric chairs? Electric chairs are horrible artifacts, but they are chairs nonetheless and finding them terrible does not change what makes a chair a chair.

Still, this move will not shield me from discussing some features that are attributed to utopianism and usually critically perceived. There are alleged attributes—such as being perfectionist—that I have to take seriously as being a possible feature of utopianism. Not committing to a definition means that I also cannot rule out that perfectionism is a part of utopian thinking and that—if perfectionism would be a problematic feature for one reason or another—it would challenge the positive value that I attribute to technological utopias insofar as they advance our understanding of the scope and location of justice.

Hence, in the following chapters, even though I will not be endorsing a definition of utopianism, I will discuss various interpretations of the general threat of perfectionism and some reasons for seeing them as less concerning. In particular, in relation to the idea that perfectionism is problematic as it requires the subjugation of diverging and equally reasonable ideas of a good society and a better way of living, which is undeniable a reality of modern societies, will necessarily require utopians to undercut peoples' freedom in the best and lead to violence in the worst case. I will argue in the following chapter that utopianism does not necessarily require the violent realization of a particular society, nor does it imply any attempt of realization—at least not in the immediate, direct way. There are other ways of exercising practical influence—both cognitive and dispositional, as I shall outline.[6] Second, I will argue that it is an integral aspect of the utopian hope that reasonable disagreement is not logically insurmountable and that critics also have not convincingly shown this and third and lastly, that it is a particular appeal of technological utopianism that it promises new solutions to contentious matters about justice in a pluralist society. This might be achieved either by envisioning new ways of realizing utopia without demanding individual changes of behavior or instantiations of new socio-political institutions; for instance, by overcoming the very sources of the disagreement (e.g. scarcity). It is a crucial part of their nature as utopias that technological utopias imagine not only a better way of living and being, but also device means to bring about social changes in a better way than other utopias or realistic approaches.

[6] I have considered using the term "directive" instead of "dispositional" for this essay. The reader shall feel free to use them interchangeably. "Dispositional" captures in my view better the contextual nature of utopias' galvanizing power. This is so, because not everyone in every situation will be enticed by utopian speculation.

Before delving deeper into the central value question as tied to notions of perfectionism, violence and stagnation (Chapters 2 and 3), however, I believe some examples of technological utopias are due. A diverse landscape of narratives about the future of society has emerged in the past decades, and increasingly more of them focus heavily on technologies and their potential transformative power (Maresch & Rötzer, 2004). One might see their prevalence as evidence against the suggestion that utopian thinking has finally become obsolete after the Fall of the Soviet Union. Though, we will discuss critics who would prefer to deny them the utopian status, as they lack some of utopias' alleged intrinsic features (collectivist spirit), those who are willing to allow some of them entrance into the traditional utopian canon suggest that in the twenty-first century "technology continues to be the primary focus of most utopian schemes, if only as principal means of bringing utopia about" (Segal, 2006, p. 75).

As I have left the contentious, definitory question about utopianism open in the previous paragraph, I chose to side with majority views concerning the question whether the following narratives deserve carrying the utopian label. The chosen examples have aroused particular attention in Utopian Studies, political philosophy or philosophy of technology and related fields and they serve my goal well to exemplify how technological utopias illuminate our knowledge and view of the scope and location of justice. Due to this already existing widespread interest in them, an abundance of secondary literature was available for me to rely on. The present essay attempts, hubristically, to reconcile insights from political philosophy, Utopian Studies and philosophy of technology, which is given already a massive undertaking given the scope of the present project. Hence, there was little room to gather own, original insights about these cases. I say this all in anticipation of the criticism that my choice represents and reproduces the aforementioned Eurocentric bias in Utopian Studies, which additionally spotlights ableist and colonial ideas—a criticism that I take serious: As I will repeat later on, I think it would be particularly promising to look, for instance, into technology-focused, anti-ableist, feminist and afrofutures to test the hypothesis of the present essay and to expand some of these ideas (Lavender, 2019; Mackey, 2009; Shew, 2020). Recent discussions of the younger Japanese utopian tradition that are heavily technology-focused also spark my interest (Ho, 1991; Moichi, 1999). A juxtaposition of these narratives with the ones I have chosen for the present analysis appears highly promising for future research.

There is another major caveat: Regarding each of the following narratives, it was not easy to distill some core elements without committing an injustice to the manifold flavors, shapes and forms in which these narratives are being presented and promoted. This is especially the case regarding the first example of a technological utopia—Transhumanism, where recent research has provided an abundance of new insights (MacFarlane, 2020). Given the present essay's focus on a methodological question, the following summaries must remain unsatisfyingly short. For a consistent presentation of those narratives within the available scope, I had to brush over various details, nuances and varieties within these movements to distill general tendencies and core ideas.

1.1 Transhumanism

It is perhaps contentious that Transhumanism would lead this list of examples of technological utopias. With Richard Saage, we will later meet a renowned utopianism scholar, who considers the Transhumanist project as a "hostile takeover" of the utopian tradition (Saage, 2013). Transhumanists themselves, including Max More, find the usage of the concept utopia as applied to Transhumanism preposterous. More refuses to attach this label, primarily because he thinks that utopias denote futuristic narratives about perfect end states and that that would immediately challenge Transhumanism:

> This criticism, and the others like it, confuse the goal of continual improvement or enhancement with the longing for a state of final perfection. These are actually radically different. The former is essentially a process of perpetual change whereas the latter is a state of stasis. Transhumanism reflects the Enlightenment commitment to meliorism and rejects all forms of apologism – the view that it is wrong for humans to attempt to alter the conditions of life for the better. Nothing about this implies that the goal is to reach a final, perfect state. (More, 2013, p. 14)

Because More thinks that Transhumanism is in search for continuous improvement and not seeking a final state of perfection—which is a sentiment shared by his ideological ancestor Julian Huxley (Huxley & Birx, 1992, p. 224)—and that such would be a feature of utopianism, Transhumanism is not utopian in his view. Of course, we are not forced to follow him here, because we have not embraced any definition of utopianism. Although, we have come across aversion of perfectionism before, I also want to postpone the substantial discussion about the problem that perfectionism poses to a later stage (Chapter 2), when we will discuss anti-utopian arguments from political philosophy. Other commentators are less hesitant to call Transhumanism utopian (Hauskeller, 2012, 2014). When Bostrom in his *Letter from Utopia* evokes the terminology, he characterizes utopia as the hope for a new and better kind of life. Thereby, echoing—knowingly or unknowingly—Ruth Levitas' definition of utopianism as a desire for a better way of living and being. He writes:

> 'Arrive?' you say; 'But isn't the journey the destination? Isn't Utopia a place that doesn't exist? And isn't the quest for Utopia, as witnessed historically, a dangerous folly and an incitement to mischief?' My friend, that is not a bad way for you to think about it. To be sure, Utopia is not a location or a form of social organization.' [...] Utopia is the hope that the scattered fragments of good that we come across from time to time in our lives can be put together, one day, to reveal the shape of a new kind of life. (Bostrom, 2008, p. 6)

What then is Transhumanism about? Transhumanism is a philosophical and intellectual movement that advocates for the use of technology to enhance and transcend the limitations of the human body and mind. Rooted in the belief that human beings can and should use scientific advancements to improve their physical and cognitive abilities, transhumanism explores the potential for radical transformations in human nature. At its core, transhumanism seeks to "free ourselves from what limits us" through the application of emerging technologies (Hauskeller, 2012, p. 39). This includes, but is not limited, to genetic engineering, nanotechnology, artificial intelligence (AI) and biotechnology. The movement envisions a future in which humans can enhance their intelligence, extend their lifespans and augment their physical capacities beyond their current, biological boundaries. One key aspect of transhumanist thought is the idea of achieving "posthuman" status, wherein individuals undergo profound alterations that go beyond typical human capabilities. This could involve integrating technological components into the human body, such as brain-machine interfaces or cybernetic enhancements, to enhance cognitive functions or physical abilities. The goal is to allow humans to transcend the limitations imposed by biology and achieve unprecedented levels of intelligence, creativity and longevity. Such ideas, as mentioned before, go back to Julian Huxley, an evolutionary humanist, who in his essay *The new divinity* does not shy away from associating his ideas with religious beliefs, considering them an "evolution-centered religion": "Thus, the central-long term concern of religion must be to promote further evolutionary improvement and to realize new possibilities; and this means greater fulfilment by more human individuals and fuller achievement by more human societies. Human potentialities constitute the world's greatest resource, but at the moment only a tiny fraction of them is being realized. The possibility of tapping and directing these vast resources of human possibility provide the religion of the future with a powerful long-term motive" (Huxley & Birx, 1992, p. 224).

Normative considerations play a crucial role in transhumanist discourse. Advocates argue that the pursuit of enhancement technologies can lead to a better quality of life, address issues of health and well-being and enable individuals to reach their full potential. Bostrom claims to be motivated by a desire "to help to make the world a better place" (nickbostrom.com). Hauskeller, again, underscores that claims such as these "reveal a conspicuous proximity to utopianism" (Hauskeller, 2012, p. 40). Critics of Transhumanism express concerns about the potential for exacerbating social inequalities, creating new forms of discrimination and raising ethical questions related to the alteration of human nature (Kass, 2003). And, naturally, the movement has been charged with naivety about the possibilities of technological progress (Peters, 2011).[7]

Transhumanism also contemplates the concept of Singularity, a hypothetical point in the future when AI surpasses human intelligence, leading to rapid and unpredictable advancements (Bostrom, 2016). Some transhumanists anticipate that the Singularity could be a transformative event, potentially enabling humans to merge with advanced AI systems or achieve a level of intelligence and understanding beyond current comprehension.

In summary, transhumanism is a forward-looking movement that promotes the use of technology to enhance and redefine what it means

[7] Ted Peters writes: "But we might ask: could progress take us to the point where a fully 'technological man' or perhaps a fully 'technologized humanity' could emerge? To believe such a thing is either possible, let alone desirable, is to embrace a myth" (2011, p. 165).

to be human. It envisions a future where individuals transcend biological limitations, achieve unprecedented levels of cognitive and physical abilities and explore new frontiers of existence. While the movement holds promise for positive transformation, it also raises ethical questions and concerns that continue to shape the ongoing discourse surrounding the intersection of technology and human enhancement. In Nick Bostrom's own words, the Transhumanist vision is "[...] in broad strokes, [...] to create the opportunity to live much longer and healthier lives, to enhance our memory and other intellectual faculties, to refine our emotional experiences and increase our subjective sense of well-being, and generally to achieve a greater degree of control over our own lives" (Bostrom, 2003, p. 494).

In the Transhumanist utopia, both of the following two narratives are oftentimes neatly integrated (Bainbridge, 2013): Through virtual realities in which one could experience various identities in the forms of virtual avatars, and in which one could experience pieces of art and experiences of utmost pleasure, virtual realities become a pre-stage playground to test humanities transformative potential. Additionally, in such virtual worlds of unlimited communication and information exchange, new forms of socializing and living emerge that promises more flourishing ways of living.

1.2 Virtual Reality

Jeron Lanier suggests that

> [virtual reality] is one of the scientific, philosophical, and technological frontiers of our era. It is a means for creating comprehensive illusions that you're in a different place, perhaps fantastical, alien environment, perhaps with a body that is far from human. And yet it's also the farthest-reaching apparatus for researching what a human being *is* in the term of cognition and perception. Never has a medium been so potent for beauty and so vulnerable to creepiness. Virtual reality will test us. It will amplify our character more than other media ever have. (Lanier, 2017, p. 5)

John Danaher considers the standard technological definition of virtual reality [VR], where it is posited "that it requires a technology that enables some degree of immersion into the simulated space. In other words, it requires a technology that creates an illusion, which can be more or less convincing, for the user that they really inhabit the computer simulation" (Danaher, 2022, p. 510). He goes on to suggest that "the distinguishing feature of this technological vision of VR is that of the computer-simulated world. It is this computer-simulated environment that provides the 'virtual' aspect of the reality inhabited by the user. With computer simulations it is, in principle, possible for people to create vast fictional worlds that are free from many of the constraints of the real world. They can also interact with other humans that share the computer simulated space or with wholly computer-programmed or artificial characters" (ibid.).

David Chalmers, a prominent philosopher, discusses the idea of *Virtual Reality+* in relation to various well-known philosophical thought-experiments that explore the nature of reality, consciousness and the implications of advanced VR technologies (Chalmers, 2022). This concept builds upon his exploration of the "Simulation Hypothesis", which suggests that our reality might be a computer simulation. In *Virtual Reality+*, Chalmers posits that as VR technology advances, it could become so sophisticated that it is indistinguishable from our physical reality. Chalmers suggests that if virtual worlds were virtually identical to our physical reality, and individuals within them had conscious experiences just like in the physical world, it would be challenging to determine which reality is the "base" reality. Chalmers argues that as VR technology becomes more immersive and capable of replicating the complexities of our sensory experiences, it raises profound questions about the nature of self and consciousness. If a person can have experiences, relationships and experiences of the same quality and intensity in a virtual world as in the physical world, does the distinction between the two realms still matter? Chalmers acknowledges that while we're not currently at a stage where virtual reality can fully replicate the richness of physical reality, technological advancements could make this a future possibility. The concept challenges us to consider the extent to which technology can influence our perceptions of reality and the self.

Chalmers' raises concerns about addiction, disconnection from the physical world and the blurring of boundaries between the real and the virtual world. While he acknowledges both risks and possible benefits of virtual reality, his normative conclusions are cautiously optimistic:

> The thesis—especially the first two parts—has practical consequences for the role of VR technology in our lives. In principle, VR can be much

> more than escapism. It can be a full-blooded environment for living a genuine life. I'm not saying that virtual worlds will be some sort of utopia. Like the internet, VR technology will almost certainly lead to awful things as well as wonderful things. It's certain to be abused. Physical reality is abused, too. Like physical reality, virtual reality has room for the full range of the human condition—the good, the bad, and the ugly. [...] But once a mature VR technology is developed, it should be able to support lives that are on a par with or even surpass life in physical reality. (Chalmers, 2022, p. xvii)

Chalmers primarily explores the implications and consequences of increasingly immersive virtual reality experiences, leaving it to individuals and society to consider the desirability of such extensions. His discussions do not take a definitive stance on whether we should or should not pursue these advancements. Instead, Chalmers prompts us to think critically about the ethical, moral and existential implications of extending virtual reality to the point where it blurs the boundaries between the virtual and physical worlds.

1.3 Computerized Communication and Cyberspace

Many hopes and many expectations have been associated with computerized communication from their early inception. Both the term "myth" and "utopia" are frequently invoked when discussing the internet and cyberspace with its various related technologies (smartphones, apps, etc.) (Kling, 1996; Mosco, 2005; Turner, 2006). The introduction of the computer as a primary means of personal and professional communication, its continued development and connection to a net of linked objects through a technology commonly described as "internet", has invoked a plethora of hopes and expectations, and summoned widespread utopia energies.

In general, to crudely combine a plethora of diverging ideas, the utopia of computerized, digital communication homes in on the widespread ignorance or *lack of information* and how they impair decision-making in various ways. Lack of information could mean that those who want to utilize the information, have no access to it—or that no one in fact has access to it. Computerization is associated with the idea that if the world were connected in the same way as a village, there would be more understanding, trust and, collective education and, therefore, improved decision-making. Such communication would be more effective, sustainable—as it would not require traveling to the sources of information and would always reach those that would also be able to make the most of them.

These and related ideas can be found in various writings, such as those of Nicholas Negroponte, Marshal McLuhan and Bill Gates among others (Daub, 2020). Analysis and criticism of these narratives have been provided by Langdon Winner and others (Winner, 1986). Natale and Ballatore summarize that understanding the "'digital sublime'- or 'cyber-utopianism' […] – has been an important topic of interest in new media studies and related disciplines. […] supporters of the digital utopia have mostly identified the internet as the carrier of a plethora of changes which should help bring about, so the argument goes, a more decentralized, democratic, and libertarian society" (Natale & Ballatore, 2014, p. 112). We will critically consider the alleged libertarianism in more detail in the next chapters. Vincent Mosco summarizes these diverse ideas as follows—a comprehensive description that is worth being fully presented here:

> The Internet provides the basis for the basis for a powerful myth because it goes a long way toward satisfying these characteristics. It is a story about how ever smaller, faster, cheaper, and better computer and communication technologies help to realize, with little effort, those seemingly impossible dreams of democracy and community with practically no pressure on the natural environment. According to this view, computer communication empowers people largely by realizing the perennial dream of philosophers and librarians: to make possible instant access to the world's store of information without requiring the time, energy and money to physically go where the information is stored. Moreover, the story continues, computer networks offer relatively inexpensive access, making possible a primary feature of democracy, that the tools necessary for empowerment are equally available to all. Furthermore, this vision of the Internet fosters community because it enables people to communicate with one another in any part of the world. As a result, existing communities are strengthened and whole new 'virtual' communities arise from the creation of networks of people share interests, commitments, and values. (Mosco, 2005, pp. 30 f.)

What such cyber-utopianism shares with Transhumanism (with the exception of Bostrom's *Fable of the Dragon Tyrant* and his *Letter from*

Utopia [Bostrom, 2005, 2008]) is that their narrative is usually not presented as a form of fictional storytelling. Most cyber-utopias are not sold as a form of entertainment literature; there is no irony to be found here, there is no constructed barrier between narrator and listener, a barrier that Thomas More in his classic *Utopia* regularly establishes and crosses regularly (Bruce, 2008). These are aspects of the playfulness of utopians: "Utopians imagine and desire radically different worlds but they often work with a light touch. They fool around with reality and tweak the nose of convention: transgressing norms, breaking rules and crossing boundaries. Utopians play with reality like a dog with a rag, twisting and shaking it until it breaks. And they poke fun; evoking satire and using jokes and wit as strategic weapons to show 'it doesn't have to be like this!'" (Sargisson, 2012, p. 16) Cyber-utopians do not weave their insights together in a *fictional narrative*, and one might, therefore, be inclined to take them at face value, as more than a mere utopian stories. This is underscored by the fact that their promises are often presented *as predictions*, made in their roles as engineers, CEOs and venture capitalists (e.g. Negroponte, Bills, Jobs and Musk).

What are the core aspects of the envisioned changes through computerized communication? Key aspects are the idea of liberation through lowering the hurdles of participating in socio-political discourses and distributing information necessary to make better informed decisions. This, as has been suggested, could lead to the empowerment of people that are suppressed by political leaders and indoctrinated through refusal of access to information (Joseph, 2012). As Natale and Ballatore write: "Digital media are embedded in a well-established, utopian narrative, by which they are represented as a liberating force, as the main actors of change within a free-market-oriented understanding of progress" (Natale & Ballatore, 2014, p. 112). The second aspect that is related to liberation of the individual is decentralization. This shall be realized by handing various policymaking processes that are currently facilitated through various models of representation directly to the people, e.g. by allowing them to directly vote on various issues, rate everything from restaurants to hotel rooms and sightseeing spots to high schools and university teachers and, thereby, use increase legitimacy and utilize "swarm intelligence". Furthermore, there is the fostering of old and the establishing of new communities. Almost all of these dimensions can be found in the works of Nicholas Negroponte and William Mitchell—to name just two notable proponents of this utopia from the mid-1990s (Mitchell, 1995; Negroponte, 1995).

One might be inclined to think that such promises that have ebbed away and resurfaced in waves since the 1960s, have undone themselves. Various authors have pointed out that none of these expectations have been fulfilled: Eva Hausteiner suggests that reality has proven those visions wrong (Hausteiner, 2022, p. 130). The internet's role in the Arab Spring is at best ambivalent—long-lasting democratic regimes have not ensued. The prevalence of face news, and the existence of filter bubbles, where people wall up to continuously reproduce evidence for their preformed judgments, indicate the vastly negative impacts of social media and forms of mass communication—people, rather than educating themselves in ways that would enhance their decision-making, numerous internet application foster that they amplify existing bad beliefs rather than revising and rejecting them. Furthermore, on a global scale, the limitations to owning and accessing digital infrastructures—the absence of electricity, money to buy and maintain the devices are significant. In recent years, it has also become more and more clear that the perception of those technologies as clean are built on an illusion (van Wynsberghe, 2021). The "dirt" that these technologies produce is effectively hidden from their users. The seeming simpleness and cleanness of pressing the "send"-button of an email program suggests that no hands have been dirtied and no actual waste has been produced. The carbon emissions, the electronic trash that evaporates poisonous gases that is killing people in the countries where the trash is "recycled", is neatly hidden behind a clean screen. Rather than providing and increasing individual freedom, various authors suggest that the ubiquity of the internet has made us pawns of an intricate version of surveillance capitalism, that commercializes and commodifies everything about us that can be turned into a data point. Shoshana Zuboff—who has underscored many of these points—spoke recently of "a death match over the politics of knowledge in the digital century. Surveillance capitalism's antidemocratic economic imperatives produce a zero-sum dynamic in which the deepening order of surveillance capitalism propagates democratic disorder and deinstitutionalization" (Zuboff, 2022, p. 2). She prompts everyone interested in fighting this match for the side of democracy to consider ways to "avert the drift into accidental dystopia" (ibid.). All of this suggests that the main elements—liberation, decentralization, empowerment and community of this vision—remain unrealized and that *the utopia of computerized communication* has failed in the face of reality. It could not and cannot deliver on its promises.

However, utopianism is not that easily refutable based on analysis or observation of reality: Reality itself is in some way a construction that lives underneath the stories we tell ourselves about it and about the future toward which we can ascend from this point in time. This concerns in particular the presence and reality of abstract political and moral concepts: Is the United States (US) a democracy? Of course, in some obvious ways, it is. In other ways, however, it is not; empirical data shows that some groups are structurally excluded from voting and the amount of people who can assume political leadership is very much confined (Dunn, 2014). Narratives must always be fundamentally different than reality, as reality will always contain more details than can be represented in a narrative. Due to the schism between reality and vision, cyber-utopians are unlikely to let go of the hopes they invest into digital communication and online technologies. The perceived unacceptability of the present will continue to fuel hopes and speculations.

References

Bainbridge, W. S. (2013). Transavatars. In M. More & V.-M. Natasha (Eds.), *The Transhumanist Reader* (pp. 91–99). Wiley.

Bostrom, N. (2003). Human Genetic Enhancements: A Transhumanist Perspective. *The Journal of Value Inquiry, 37*(4), 493–506. https://doi.org/10.1023/B:INQU.0000019037.67783.d5

Bostrom, N. (2005). The Fable of the Dragon Tyrant. *Journal of Medical Ethics, 31*(5), 273–277. https://doi.org/10.1136/jme.2004.009035

Bostrom, N. (2008). Letter from Utopia. *Studies in Ethics, Law, and Technology, 2*(1), 1–7. http://www.nickbostrom.com/utopia.html

Bostrom, N. (2016). *Superintelligence: Paths, Dangers, Strategies*. Oxford University Press.

Bruce, S. (2008). Introduction. In S. Bruce (Ed.), *Three Early Modern Utopias* (pp. ix–lxi). Oxford University Press.

Chalmers, D. J. (2022). *Reality+: Virtual Worlds and the Problems of Philosophy* (1st ed.). W. W. Norton.

Danaher, J. (2022). Virtual Reality and the Meaning of Life. In I. Landau (Ed.), *The Oxford Handbook of Meaning in Life* (pp. 508–524). Oxford University Press.

Daub, A. (2020). *What Tech Calls Thinking—An Inquiry into the Intellectual Bedrock of Silicon Valley*. Farrar.

Dunn, J. (2014). *Breaking Democracy's Spell*. Yale University Press. http://www.jstor.org/stable/j.ctt1bhkp0j

Estlund, D. (2020). *Utopophobia—On the Limits (If Any) of Political Philosophy*. Princeton University Press.

Hauskeller, M. (2012). Reinventing Cockaigne: Utopian Themes in Transhumanist Thought. *Hastings Center Report, 42*(2), 39–47. https://doi.org/10.1002/hast.18

Hauskeller, M. (2014). Utopia. In S. L. Sorgner & R. Ranisch (Eds.), *Post- and Transhumanism: An Introduction* (pp. 101–108). Peter Lang.

Hausteiner, E. (2022). Radikale Netzwerke. Über die Notwendigkeit digitaler Utopien. In B. Klein & R. Schmidt (Eds.), *Was macht die Digitalisierung mit der Politik?* (pp. 123–136). De Gruyter. https://doi.org/10.1515/9783110785265-010

Ho, K.-k. (1991). Japanese in Search of Happiness: A Survey of the Utopian Tradition in Japan. *Oriens Extremus, 34*(1/2), 201–214. http://www.jstor.org/stable/24047174

Huxley, J., & Birx, H. J. (Eds.). (1992). *Evolutionary Humanism*. Prometheus Books.

Jameson, F. (2005). *Archeologies of the future: The desire called utopia and other science fictions*. Verso.

Joseph, S. (2012). Social Media, Political Change, and Human Rights. *Boston College International and Comparative Law Review, 35*(1), 145–188.

Kass, L. (2003). Ageless Bodies, Happy Souls Biotechnology and the pursuit of perfection *The New Atlantis* (1), 9–28. https://www.thenewatlantis.com/publications/ageless-bodies-happy-souls

Kling, R. (1996). Hopes and Horrors: Technological Utopianism and Anti-utopianism in Narratives of Computerization. In R. Kling (Ed.), *Computerization and Controversy—Value Conflicts and Social Choices* (2nd ed., pp. 40–58). Morgen Kaufmann Publishers.

Kumar, K. (1991). *Utopianism*. Open University Press.

Lanier, J. (2017). *The Dawn of the New Everything: Encounters with Reality and Virtual Reality*. Henry Holt and Company.

Lavender, I. (2019). *Afrofuturism Rising: The Literary Prehistory of a Movement*. Ohio State University Press. http://www.jstor.org/stable/j.ctv2zbhw62

Levitas, R. (2011). *The Concept of Utopia*. Peter Lang.

Levy, J. T. (2016). There Is No Such Thing as Ideal Theory. *Social Philosophy and Policy, 33*(1–2), 312–333. https://doi.org/10.1017/S026505251600025X

Lillehammer, H. (2011). The Epistemology of Ethical Intuitions. *Philosophy, 86*(336), 175–200. http://www.jstor.org/stable/23014834

Longxi, Z. (2002). The Utopian Vision, East and West. *Utopian Studies, 13*(1), 1–20. http://www.jstor.org/stable/20718406

MacFarlane, J. M. (2020). *Transhumanism as a New Social Movement: The Techno-Centred Imagination*. Palgrave (Springer).
Mackey, P. J. (2009). Crip Utopia and the Future of Disability. *Critical Disability Discourses, 1*. https://cdd.journals.yorku.ca/index.php/cdd/article/view/23383
Maresch, R., & Rötzer, F. (Eds.). (2004). *Rennaisance der Utopie - Zukunftsfiguren des 21. Jahrhunderts*. Suhrkamp.
Mitchell, W. J. (1995). *City of Bits: Space, Place, and the Infobahn*. MIT Press.
Moichi, Y. (1999). Japanese Utopian Literature from the 1870s to the Present and the Influence of Western Utopianism. *Utopian Studies, 10*(2), 89–97. http://www.jstor.org/stable/20718096
More, M. (2013). The philosophy of Transhumanism. In M. More & N. Vita-More (Eds.), *The Transhumanist Reader* (pp. 3–17). Wiley-Blackwell.
Mosco, V. (2005). *The Digital Sublime: Myth, Power, and Cyberspace*. MIT Press.
Natale, S., & Ballatore, A. (2014). The Web Will Kill Them All: New Media, Digital Utopia, and Political Struggle in the Italian 5-Star Movement. *Media, Culture & Society, 36*(1), 105–121. https://doi.org/10.1177/0163443713511902
Negroponte, N. (1995). *Being Digital*. Knopf.
Peters, T. (2011). Transhumanism and the Posthuman Future: Will Technological Progress Get Us There? In W. Grassie et al. (Eds.), *H+/-: Transhumanism and Its Critics* (pp. 147–175). Metanexus.
Popper, K. R. (2012). *The Poverty of Historicism* (2nd ed.). Taylor and Francis.
Saage, R. (2013). New Man in Utopian and Transhumanist Perspective. *European Journal of Futures Research, 1*(1), 14. https://doi.org/10.1007/s40309-013-0014-5
Sargent, L. T. (1994). The Three Faces of Utopianism Revisited. *Utopian Studies, 5*(1), 1–37. http://www.jstor.org/stable/20719246
Sargent, L. T. (2010). *Utopianism: A Very Short Introduction*. Oxford University Press.
Sargisson, L. (2012). *Fool's Gold?: Utopianism in the Twenty-First Century*. Palgrave Macmillan. https://doi.org/10.1057/9781137031075
Segal, H. P. (2006). *Technology and Utopia*. American Historical Association.
Segal, H. P. (2012). *Utopias: A Brief History from Ancient Writings to Virtual Communities*. Wiley.
Shew, A. (2020). Ableism, Technoableism, and Future AI. *IEEE Technology and Society Magazine, 39*, 40–85. https://doi.org/10.1109/MTS.2020.2967492
Thaler, M. (2018). Hope Abjuring Hope: On the Place of Utopia in Realist Political Theory. *Political Theory, 46*(5), 671–697. https://doi.org/10.1177/0090591717740324

Turner, F. (2006). *From Counterculture to Cyberculture: Steward Brand, the Whole Earth Network, and the Rise of Digital Utopianism*. The University of Chicago Press.

van Wynsberghe, A. (2021). Sustainable AI: AI for Sustainability and the Sustainability of AI. *AI and Ethics, 1*(3), 213–218. https://doi.org/10.1007/s43681-021-00043-6

Winner, L. (1986). Mythinformation. In *The Whale and the Reactor* (pp. 98–117). University of Chicago Press.

Zuboff, S. (2022). Surveillance Capitalism or Democracy? The Death Match of Institutional Orders and the Politics of Knowledge in Our Information Civilization. *Organization Theory, 3*(3). https://doi.org/10.1177/26317877221129290

CHAPTER 2

Perfectionism, Stagnation and Transcendental Theorizing

Abstract Various arguments that consider utopianisms as risky or dangerous will be discussed—many of them related to the notion of perfectionism: violence as a result of careless politics, stagnation and the assumption of normative superiority. Rebutting these concerns, paves the way toward a more positive perception of utopianism. Utopianism, while unable to directly guide governance, can fulfill both important cognitive and dispositional functions, which are intertwined.

Keywords Violence · Karl Popper · Perfectionism · Stagnation · Fallacy of approximation · Cognitive function · Dispositional function · Estrangement · Critique

Having introduced some paradigmatic technological utopias briefly in the last chapter, I will now turn toward surveying some of the general criticism of utopianism that undergird the anti-utopian position. This shall eradicate some doubts regarding some alleged dangers and problems. As one of the most prominent anti-utopians, we shall commence by analyzing Karl Popper's famous charges against utopianism and unpack what they mean for current technological utopianism. Each of the following charges are in a fuzzy way related to the idea of fatal perfectionism. This, as we shall see, though, is an ambiguous notion and one can understand the associated problem in various ways, each demanding

M. Sand, *Technological Utopianism and the Idea of Justice*,
https://doi.org/10.1007/978-3-031-75945-1_2

a different response. We shall see that neither political, nor technological utopianism entails violence, nor even necessarily the impetus for realization. Utopia can be practical in ways that are not confined to steering current policymaking (Sect. 2.2). Additionally, we will see that the threat of boredom and stagnation is not straightforward either. I will argue that stagnation out of a loss of inclination, perhaps due to voluntary choices—or, in fact, because all goods have already been achieved and no objectively important activities remain—is a morally more innocuous form of stagnation than that which is (violently) imposed (Sect. 2.3). Lastly, in Sect. 2.4, I will suggest that even if utopianism were in some sense perfectionist (adopting a stance of normative superiority or epistemic absolutism), this is not unique to the utopian outlook and lacks immediate threat.

Hans-Peter Schütt suggests that the anti-utopian impulse stems from an aversion of the idea that people, their needs, inclinations and behavior become part of backing recipe in the utopian imagination (Schütt, 2010). In this view, the utopian sees the individual as an item on a list of ingredients that need to be carefully stirred, mixed and put into a form and then into the oven; a powerful metaphor that raises suspicion. Why exactly, however, would forming a part in a baking *recipe* be morally problematic? If the recipe remains entirely conjured in the imagination of someone, it does not seem to be quite problematic. It neither harms anyone nor does it have any effects at all.

Properly understood, thus, the charge is better directed at the idea of blueprinting the perfect state, which then serves as a model according to which real societal and political interventions *shall be executed*. Though even this idea is highly ambivalent, and it is not apparently obvious when stated as such, what exactly is wrong with it. We shall unpack this in more detail in the present chapter. One of the utopian retreats against this charge is to suggest that utopians—at least postmodern ones—refrain from such kind of perfectionism and blueprinting. This move is initially akin to the previously discussed strategy of defining blueprinting out of the scope of the concept of utopia and simply declare them to be something else, namely blueprints instead of utopias.

Sargent, for instance, suggests that blueprinting is not the utopians' business anymore siding with what "[…] the American political theorist George Kateb […] wrote [that] 'any serious utopian thinker will be made uncomfortable by the very idea of blueprint, of detailed recommendations concerning every facet of life'. In other words, the utopian argument is that utopias do not create the artefacts that Popper and others say they do" (Sargent, 2010, p. 96). Popper's name is here invoked as a precursor of the general idea that seems to underlie Schütt's argument and he—as we will see shortly—delivers it with much more normative impetus. Sargisson suggests in similar fashion, as we have seen before, that "[s]ome people do sometimes try to realize their utopia [...]. But realization is not part of my definition of utopia. Many utopias exist as thought experiments and it would be a mistake [...] to define them as realizable blueprints or road maps to the good life" (Sargisson, 2012, p. 15).

Unlike Sargisson, I am not bound to a definition of utopianism. Unfortunately, that also means that I cannot retreat to these definitory restrictions that Sargent and Sargisson propose. Perhaps, perfectionism with the various risks it contains, is present in some form or another in utopias and in technological utopias, in particular. This possibility cannot be readily dismissed. The Transhumanist idea that human life could be so much better, experientially and morally, while perhaps not ever becoming perfect as Max More suggests (More, 2013), still entails the value judgment that the present way of being is highly defective and better ways of being are conceivable, achievable and desirable. While this might not entail perfectionism in one sense of the term—static, end-state perfectionism, as one might call it—it likely entails the belief in the normative superiority of the value judgment—namely that Transhumanism is *the best answer* to certain perceived imperfections and deficiencies, in which every way one should try to overcome them. This stance of value superiority might be considered as yet another form of perfectionism. Thus, perhaps perfectionism *is* present in one form or another in many utopias and in particular technological utopias. The idea seems not far-fetched, when reading passages such as these from Negroponte's *Being Digital*: "But more than anything, my optimism comes from the empowering nature of being digital. [...]. The information superhighway may be mostly hype today, but it is an understatement about tomorrow. It will exist beyond people's wildest predictions. As children appropriate a global information resource, and as they discover that only adults need learners permits, we are bound to find new hope and dignity in places where very little existed before" (Negroponte, 1995, p. 231).

Rather than asking, however, whether perfectionism is *integral* to utopianism, which would bring us closer to the definitory question that I attempted to avoid in the previous chapter, is the question: If they were feature of utopianism, would blueprinting and perfectionism really be that problematic and, if so, in all its different forms? The problem of blueprinting cannot be one as such—as I have suggested before vis-á-vis Schütt's recipe metaphor: Why would one be offended for being part of a recipe that is never baked? There seems little to no damage here worth a complaint. It is rather the *attempt* to bake the cake according to the recipe that seems to underlie the problem that many anti-utopians since Popper perceive. Let us take this as the first interpretation of the charge of perfectionism and unpack its normative significance. The criticism could be stated as follows: Utopian perfectionism and blueprinting suggest that whole societies can be envisioned and predicted. This is on the one hand epistemically unfeasible. Furthermore, the practical goal of remodeling a society comprehensively is doomed to lead to violence and failure.

2.1 The Failures of Predicting and Remodeling Society

Each of the aforementioned assumptions can be found prominently in the works of Karl Popper, whose views were briefly introduced before. We have seen that Popper conceptualizes utopianism as the idea of a comprehensive reconstruction of society and not only as a creative, intellectual exercise. He thinks that the utopian project commences with the intent to interfere in a comprehensive way with society, which is both morally and epistemically problematic. Popper claims that historicists, holistic state planners and utopians form an unholy alliance in their attempt to both predict the consequences of certain comprehensive interventions into society *and* then execute those interventions at all costs—although the predictive aspect is less articulate in many forms of utopianism, the interventionist paradigm makes utopians as dangerous as their historicist counterparts. He suggests that the blueprints utopians conjure will demand that "we 'mould' […] men and women to fit into this new society" (Popper, 2012, p. 64). In this terminology, the fear of becoming part of a baking recipe suddenly gains more normative traction. According to this view, an individual's life with one's personal desires and preferences does not only appear as an item on a cardboard; that life really gets mixed and mingled, formed and baked in the way the utopian envisions. Without even using the terminology, one senses immediate concerns about peoples' liberty; that freedom will be undermined, and people not only stimulated or nudged, but literally coerced and forced into a new way of living. Popper suggests that both the historicist and the utopian "attempt to rationalize this change, the one by prophesying the course of social development, and the other by insisting that the change should be strictly and completely controlled, or even that it should be entirely arrested. The control must be complete, for in any department of social life which is not so controlled, there may lurk the dangerous forces that makes for unforeseen changes" (ibid., p. 68). We see here that Popper indeed believes that utopian remodeling of society presumes the effort *to predict and to reshape* society on a grand scale. An intellectual who according to Popper embodies the alliance of historicism and utopianism is Karl Mannheim.

We will return to the issue of prediction in more detail later on (Sect. 3.2) and see that we are not forced to read technological utopias as predictions. In Negroponte, we encountered a technological utopian, who indeed fancies predictions. Ray Kurzweil's *The Age of Spiritual Machines* is replete with predictions, for instance, that around the year 2029 "the life expectancy of humans continues to increase and is now around 120 years" (Kurzweil, 1999, p. 223). However, we must contrast the nature of these instances of prediction with historicism, whose underlying scientific paradigm that understands wider societal change as being based on some law-like regularities that allow for predictions, has largely disappeared from the social sciences. Furthermore, the mentioned Negroponte and Kurzweil seem to constitute a rarity among the technological utopians. Many of them are hesitant to confidently speak about the possible consequences that the envisioned interventions will have. Transhumanist Max More's stance might be more exemplary, who suggests that "[no] specific predictions, however, are essential to transhumanism. Transhumanism is defined by its commitment to shaping fundamentally better futures as defined by values, goals, and – general direction, not specific goals. Even to the extent that a goal is somewhat specific – say, abolishing aging, becoming post-biological, or enhancing cognitive abilities to some arbitrary degree – the means and time frame in which these might be achieved are open to differing views. Transhumanism per se says much about goals but nothing about specific means or schedules" (More, 2013, p. 28). Though, the existence of predictions associated with technological utopias does *not* force us to *reduce* their value to that feature.

Later, it becomes clear that Popper's resistance toward utopianism is more directed against its alleged impulse of reshaping or remodeling society—and we will see that he has a specific style or mode of remodeling in mind that he attributes to his image of the utopian planner.

This remodeling must, as Popper puts it most explicitly in his later work, culminate in coercion and violence:

> That the Utopian method, which chooses an ideal state of society as the aim which all our political actions should serve, is likely to produce violence can be shown thus. Since we cannot determine the ultimate ends of political actions scientifically, or by purely rational methods, differences of opinion concerning what the ideal state should be like cannot always be smoothed out by the method of argument. They will at least partly have the character of religious differences. And there can be no tolerance between these different Utopian religions. Utopian aims are designed to serve as a basis for rational political action and discussion, and such action appears to be possible only if the aim is definitely decided upon. Thus the Utopianist must win over, or else crush, his Utopianist competitors who do not share his own Utopian aims and who do not profess his own Utopianist religion. (Popper, 1986, p. 6)

Popper himself has experience and lived through the age of political extremes. He saw the rise of Nazism in Germany and Austria and had to flee to the United Kingdom (UK). He lost family members being killed for their religious-ethnical roots. He flirted with communism, but also saw how leftists violently mobbed through the streets of post-war Vienna in 1919. These experiences will readily explain his sensibilities and aversions and fueled his views on utopian speculation. Still, we must ask the question, how warranted are those views? Does he not dangerously conflate the much wider concept of utopianism with the totalitarian ideologies of the early twentieth century? It seems so.

Popper's views invite two distinct responses. First, I will suggest that the view falsely underestimates the danger of a more careful mode of policymaking (piecemeal engineering, as Popper calls it), ignoring the violent history that many liberal states have been built upon, and the "silent" violence that they seem willing to accept in clinching to the ideal of modest political progress. Here and in other parts of the present essay, I shall employ the following paradigm: We must consider objections toward a certain way of theorizing about technological and socio-technical change (e.g. more or less realistic) in light of the alternative views that are on offer. In this manner, I will regularly press the reader to consider whether the criticism of (technological) utopianism is unique to the utopian outlook, or rather poses problems for any ways of theorizing about society and politics, including those that are deemed more realistic and liberal.

Second, I will suggest that the view rests on the mistaken belief—although not quite unreasonable when looking into the context of technological utopia—that utopianism is necessarily a practical endeavor. While it is true for some (technological) utopians, who pursue through individual modification, community building and advocacy changes toward utopia (see Sect. 3.1), it is not necessarily integral to utopianism that it ought ever to become practical. Even if it were a practical endeavor, the way it commences is not necessarily in the way Popper suggests—it is not necessarily violent and coercive.

Let us start with the first point: Popper suggests, as we saw before, that utopianism always has a "public character" (2012, p. 61), which aims at "remodelling the 'whole of society' in accordance with a definitive plan or blueprint [...]". But such "remodeling of the whole" seems ambivalent: I said before that Popper dangerously equates the remodeling of the "whole of society"—which could be better termed *scope* of remodeling—with the *depth* of the remodeling. Neither, as I will argue, can function to sufficiently distinguish utopian from other—more realistic or liberal—forms of political intervention.

We can readily see that with regard to scope—understood as socio-political interventions that affect *everyone to some degree*—piecemeal engineering and more "modest" political reform is not free of interventions with great scope. Take, for example, the realistic political intervention of introducing the general duty for men and women to undergo military service (conscription). Compulsory military service in Germany has been abolished only in 2015. In other democratic states, however, for both men and women military service remains obligatory. Under the pressure that the ongoing war in Ukraine has created, Germany is currently discussing the reintroduction of conscription. Hence, it should be clear that this is by no means a utopian political proposal. However, its scope is wide: The administrative costs are enormous, and every citizen of a certain age will be affected to a significant degree. Hence, we see that also realistic political reform has far-reaching implications for citizens and their liberties. Remodeling a society in such piecemeal fashion can, therefore, also have significant *scope*. Now, remodeling in terms of depth means that people are *affected in a fundamental sense*. One could say it affects *central values* or convictions. That does not have to be the same as affecting everyone. The abolishment of the property of big landowners is perhaps such intervention that does *not* affect everyone, but those who are affected by it, are affected in a very *fundamental way*.

Perhaps, it is such interventions in terms of depth that makes utopian practice so highly problematic and that distinguishes the utopian from a piecemeal engineer—Popper's opponent to the utopian; his version of political leader, who is more liberally minded, scientifically guided and, in her political actions aware of her epistemic shortcomings. However, are those interventions distinguishable in terms of scope and depth—and can some form of piecemeal engineering really *avoid* interventions that have substantial scope and depth? The example of conscription seems instructive again: The intervention does seem to touch a very fundamental value—that of the liberty of all individuals of a certain age. Hence, it seems increasingly doubtful that depth and scope of remodeling can successfully demarcate utopian from other forms of political interventions. Initially, Popper insinuates that there is a difference between political interventions in terms of how holistic the political interventions are—roughly coinciding with my previously introduced concept of scope: The piecemeal engineer "tries to achieve [some ideals] by **small adjustments** and re-adjustments which can be continually improved upon. [...] he will avoid undertaking reforms of complexity and scope which make it impossible for him to disentangle causes and effects, and to know what he is really doing" (Popper, 2012, p. 61; own emphasis). Here, Popper equates scale of intervention with the "smallness of adjustments". This, as we have seen, is ambivalent and could be further refined into the question of scope and depth, as introduced above. Furthermore, he adds that continuously accumulating "small adjustment" could—in the long run—lead to very fundamental societal transformations: "[...] nor shall I exclude the possibility that a series of piecemeal reforms might be inspired by one general tendency, for example, a tendency towards a great equalization of incomes" (p. 62). So, he goes on to admit that he "shall not attempt to draw a precise line of demarcation between the two methods [holistic vs. small scale]" but rather moves on to distinguish the two via an analysis of their affinity to risk and their willingness to consider failure and reversibility of their interventions, or so it seems. He seems to agree with the impossibility of distinguishing utopian politics from piecemeal engineering by way of scale and scope, which seem equally indistinguishable, or scale and depth when writing: "[...] the difference between Utopian and piecemeal engineering turns out, in practice, to be a difference not so much in scale and scope as in caution and preparedness for unavoidable surprises" (Popper, 2012, p. 63). Unlike the piecemeal engineer, which the utopian rejects for being "too modest", they "always fall back on somewhat haphazard and clumsy although ambitious and ruthless application of what is essentially a piecemeal method without its cautious and self-critical character" (p. 63).

This is the crux of Popper's argument: What he identifies as key diverging characteristic of these different forms of political practice are rather matters of temperament and virtues that lead to different *modus operandi* rather than substantive differences in terms of scope and depth of their political interventions. But can that distinction between a careful, piecemeal approach and a reckless utopian social planning approach be consistently upheld?

This seems rather unlikely: It is hard to ignore that many liberal states have historically employed interventions that have proven to be rather irreversible and incautious. This is most obviously the case, when it comes to the foundational moments of many current liberal states (France and the US) and their responses to threats against their sovereignty. Barbara Goodwin writes: "All social reform displeases some vested interests and is carried through by legal coercion: the utopian need be no more coercive than were the liberal founders of the welfare state" (Goodwin, 1980, p. 396). Slavoj Žižek reminds us vividly of the violence that the "exportation" of liberal democracies have brought to the Middle East as a response to the Terrorist attack on the World Trade Center that—in their views—seemed to threaten their sovereignty (Žižek, 2008). While Popper suggests that piecemeal engineering proceeds more cautiously with foresight and aware of their fallibility (ibid., p. 63), these historical examples show that liberal democracies—and not only totalitarian regimes that have committed themselves to some superficial alliance with utopianism—continuously employ interventions with massive scope (and depth) that can lead to chaos and irreversible conflicts.

These examples still seem to suggest that those are the consequences of a version of extremist liberal politics that have ultimately left their ideological roots (the piecemeal approach) and have become totalitarian and more akin to some utopian schemes and modes of political intervention. As a response, Popper could claim that these examples are not representative of piecemeal social engineering as he understands it. Such interpretation seems plausible enough, yet invites two important responses: First, in this light, is piecemeal engineering then not in itself a political ideal that remains seemingly unrealized and perhaps entirely unrealistic? Has it ever been and can it actually take off as a mode of governance without ruthlessly and uncompromisingly ridding itself initially of staunchest political opposition (as happened in France, the US and other liberal democracies that emerged out of wars and civil wars).

And, more importantly, is it even desirable that political leaders always proceed in a stance that is reversible, cautious and modest? This is to be doubted: I suggest that piecemeal engineering as sketched before, has not only historically never been consistently upheld, but it should also and cannot consistently be morally afforded: For any reasonable practice-oriented political approach, there can be circumstances that morally demand radical interventions in terms of scope and depth into society.

This is an important insight that deserves closer inspection. Whether a radical, potentially coercive political intervention—an intervention of great depth so to speak—is deemed morally or politically acceptable is not only a matter of fancy of the political leader. Popper suggests that the piecemeal engineer is at liberty to remain inactive or at least highly cautious in situations in which he is lacking sufficient insight and will, therefore, proceed with utmost care: "The piecemeal engineer knows [...] how little he knows. He knows that we can learn only from our mistakes. Accordingly, he will make his way, step by step, carefully comparing the results expected with the result achieved, and always on the look-out for the unavoidable unwanted consequences of any reform [...]" (2012, p. 61). To me, this presumes a privileged political position that is entirely absent before the emergence of a more or less stable society, which is rare now and almost impossible to sustain. The possibility of political inaction, cool and caution are morally and politically acceptable stances only when there already is a baseline of welfare, security and sovereignty.

Sustaining or even being in such politically privileged position, however, is by no means within the piecemeal engineers' control. The need for political action of massive scope or depth can arise due to exogenous causes of various kinds—man-made and natural; attacks by other states that threaten the sovereignty of certain states (Ukraine), the dramatically increasing consequences of climate change, natural catastrophes (pandemics such as COVID-19), to name but a few. The portrayal of the utopian as someone, who summons radical change merely out of a fancy of idealisms, underestimates the perceived moral need to respond adequately to such varied exogenous causes that threaten even the most minimal aspirations of any political body.

Under any circumstances, it must remain the piecemeal engineers desire to ensure his citizens' health, the continuous sovereignty of the political body (the state), some range of liberty of his citizens and the possibility for them to live decent lives—or some such political ambition. It is a major neglect that Popper continuously talks about the methods and modes of political intervention rather than the goals of policymaking. Only briefly does he remark that the piecemeal engineers' "ends may be of diverse kinds, for example, the accumulation of wealth or of power by certain individuals, or by certain groups; or the distribution of wealth and power; or the protection of certain 'rights' of individuals, or by certain groups, etc." (Popper, 2012, p. 61).[1] Political leaders must have at least such minimal goals, otherwise any method or mode of intervention is *directionless*. Now, importantly, in times of crisis and despair due to manifold causes, even these minimal goals of "protecting certain 'rights' of individuals" suddenly become highly idealistic ambitions that might require rather radical measures to be achieved.

In short, whether caution or recklessness are the *right* modes of political intervention, is not an intrinsic feature of these modes of doing politics. Just like the virtues of carefulness and modesty are misplaced in situations of despair and dire need, so is piecemeal engineering in situations of political crisis and threat. The different modes of political inference that Popper sketches and their perhaps divergent impact on peoples' lives and their liberty, have no fixed normative nature: They are neither categorically good nor bad. Rather, it is reasonable to assume that their political or moral rightness falls within a continuum, depending on how worse the current state of affairs is. Liam Murphy has suggested in his *Ideal Theory in a Non-Ideal World* that the classical charge of demandingness against the principle of beneficence would have to be expanded to all moral theories (Murphy, 2000): All moral theories will reasonably demand much sacrifice from individuals in times of despair. An

[1] Although, often invoked, it is quite misleading given this vast array of different political goals that might be approached via piecemeal engineering, to call Popper a liberal or even libertarian—as is often done (Goodwin, 1980). Popper himself suggests rather oddly that "[p]ublic or political social engineering may have the most diverse tendencies, **totalitarian** as well as liberal" (2012, p. 61; own emphasis). Although between the lines, an appeal to liberalism shines through his works, expressions such as these suggest that the typical distinctions between political theories—liberal, republican, socialist, fascist, etc.—run diagonal to Popper's introduced distinction between piecemeal engineering and utopian planning.

analogous argument has just been presented here: The worse the actual socio-political state is, the more substantive—both in scope and depth—a political intervention will be required to fix the evils of the current state of affairs. This does not mean that violent intervention *is necessary*. But to *insist*—when finding oneself in a state of widespread despair and misery—that something less than far-fetched speculation and more fundamental change of the political reality is needed and to rather demand careful, piecemeal change—means to succumb to desperation. This cannot be politically or morally acceptable either.

This points to an important insight that I shall repeat later: Whether or not utopian speculation is necessary—whether substantial reform (in whichever mode pursued) seems somehow at least worthy of being envisioned, summoned and contested, also depends on how inacceptable the world and our current state of being are: Much more than being associated with an optimistic stance about the possibility of *progress in the future*, as utopianism is often understood, it might rather be associated with a massive degree of dissatisfaction and upset about our *current state of being* that induces the desire for change (Mao, 2020). Climate change, racial and gender discrimination, and growing global inequalities are perceived as realities that spur political imagination and demand fundamental changes to how we live (Thaler, 2022). In light of such realities, a careful and piecemeal approach must be seen as complicit with the cruelties of the *status quo* in the eyes of those who suffer under said realities.

In the meantime, we have paused a question that has been raised before and to which we must return now: Does the utopian necessarily have to succumb to violence when realizing her vision? Consider, the long-held belief—already found in More's *Utopia*—that property is corrupting societies and ultimately undermining the happiness of their members. Popper might be inclined to think that a utopian, who fancies a society without private property, cannot count on the willingness of property owners to abandon their estates, etc. Hence, she will have to move forward seizing property and must expect resistance (as happened in the Bolshevik revolution in Russia). Popper can be understood in two ways: First, he might deem the seizing of property itself conceptually as an act of violence, as it demarcates a political intervention of enormous *depth*—as we have previously characterized it; it concerns important values and convictions of the members of society. In a second reading, he might think that such acts necessarily lead to violence, as property owners can naturally be assumed to resist. Although, it is beyond the scope of the present essay to discuss this in detail, it should be clear that neither of these interpretations is obviously true. Even regarding property, there can be non-violent ways of redistribution and revocation that have historically been accepted without violent response (e.g. taxation [Goodwin, 2008]) and whether such interventions undermine certain values (e.g. liberty) in a fundamental way, of course, requires a much larger debate that must consider the genealogy of said property distribution. While some utopian scholars warn of the risks of blindly implementing utopian schemes and taking them too much at face value, Popper's straightforward identification of utopianism with political violence has almost unequivocally been rejected and should require no further inspection (Goodwin, 1980); it is both conceptually and empirically awry.

More interesting and fundamental is the following question: Does the utopian actually need to act on her conviction about property at all? In which way, if any, is utopianism practical? Popper's criticism rests on the assumption that utopianism is a practical endeavor of a particularly narrow kind. Utopia is seen as the blueprint that assertive political leaders strive to realize. This becomes explicit in another passage, where Popper writes: "Utopian aims are designed to serve as a basis for rational political action [...]" (Popper, 1986, p. 6). Various commentators and historical examples suggest the opposite. While, sure enough, utopians such as Fourier and Owen have tried to develop detailed blueprints with the intent of realization, many utopian scholars suggest that both traditional utopias and many modern ones "do not consider themselves as the destination of history; rather, they are notional alternatives to the undesirable development of their societies of origin" (Saage, 2013, p. 5). Richard Saage suggests that the functional outlook of utopia changes in the early period of industrialization. We will see later that he and many others are more reasonably concerned that contemporary technological utopias have become quite a bit more practical nowadays than their traditional precursors (Saage, 2019). Many technological utopians while staying within the confines of democratically legitimate activities assume a more proactive stance. They forge alliances, advocate and lobby, investigate, model and engineer to further their visions.

Still, this is neither the realm of policymaking narrowly conceived (governance), nor representative of the utopian tradition. Krishan Kumar makes the same objection not regarding the semantics of the term "utopia", but with regard to its history. He shows that many utopians were not interested in the implementation of their ideas in their respective contemporary society (see also (Sargisson, 2012, p. 15). The fact that utopias are not realized nor intended to be realized does ***not imply*** that they have no ***practical value***—they can still have various positive effects, motivational, creative, inspirational and regulative. We shall consider them as the cognitive and dispositional functions of utopianism in a moment. Kumar writes:

> We are forced, once again, to consider that utopian thought and utopian practice may be different things, not to be judged by some presumed correspondence between them. [...] Their ideal of perfection is theoretical; their writers may be quite indifferent to the problems of achieving their goal in practice (which is not the same thing as saying that they are indifferent to the practical value of their utopias). (Kumar, 1991, pp. 72 f.)

2.2 Varieties of Practical Relevance

We must agree with Popper that one form of practical orientation would make the utopian impulse indeed dangerous: If utopianism were a project of concocting political "recipes" (as Schütt suggests) ***and*** then attempting to bake the political dish of a new society based on said recipes, one would have reason to worry. But, as we have seen, utopians rarely operate in that mode of devising recipes and trying to cook them up. David Estlund has, therefore, starkly made clear "[...] that the correct (or true, or best) theory of justice might be without practical relevance" (Estlund, 2020, p. 304). This statement, however, is equally in need of qualification. We have hereby arrived at a rather dissatisfying dichotomy: It seems that we must either assume that utopias are entirely without practical relevance (Estlund), or always on the lookout for practical realization with the consequence of coercion and violence (Popper).

I suggested before that technological utopias as a subset of utopias within the wider tradition are worthwhile to engage with. This provokes the question: If it is not practical relevance of the Popperian type, what then makes (technological) utopianism valuable and worthwhile of engagement? There are two possibilities that we shall briefly discuss. The first, is the suggestion by David Estlund, that ideal theories of justice—the type of utopias predominantly discussed in current political philosophy—might be entirely without practical value, but nevertheless ***intrinsically*** valuable. As it remains unclear just to how much value the intrinsic value of utopias amount to, however, I shall argue that the view cannot support my position of worthwhile engagement. A second route that I deem more promising, endorses the ideas of some prominent scholars from the field of Utopian Studies. Their suggestions—too manifold and rich to be comprehensively discussed here—include, among others, that utopias' practical relevance is rooted in its cognitive power: The envisioning of ways of living and being better entices reflection and challenges our various ideological imprisonments. They estrange us from present socio-political realities and ways of living and, thereby, induce reflection and a more critical perspective on aspects of our surroundings that we might have previously considered immutable. Various versions of this basic idea of utopias' cognitive relevance have been advanced and promoted in the field of Utopian Studies. To me, these ideas seem highly plausible, and I will rather uncritically adopt them. Later on, I will delineate in a more fine-grained way how exactly the critical, subversive function of utopianism resurfaces in the context of technological utopias (Chapters 4 and 5).

Recent political philosophy has discussed the concept of utopianism primarily in relation to theories of justice that commence presuming rather unrealistic conditions. Rawls' theory of justice usually serves as a prime example: It is designed under two assumptions that are not and perhaps never will be reality; they are seen—in other words—as being idealistic or utopian. Those two assumptions are that (i) all relevant agents comply with the demands of justice applying to them; and that (ii) natural and historical conditions are favorable (Rawls, 1999; Valentini, 2012). Through the works of political philosophers such as Estlund and others, we see that the confinement of a discussion about utopianism to theories of justice with the aim of justifying principles of justice, have proven to be analytically highly fruitful. He and other authors have contributed to a richer understanding of political feasibility, problems of rational choice, social and political progress and other related topics. Despite this analytical fruitfulness, it is unfortunate for reasons that shall become increasingly clear that traditional utopias have, thereby, been pushed out of the purview of this discussion. The creative input that traditional and modern utopias could contribute to those discussions is underestimated.

Estlund arrives at his conclusion about utopias' utter impracticality accepting that it is perhaps true that some conditions toward an ideal state are so interdependent that they need to be conjointly fulfilled to approximate an ideal state. Trying to fulfill certain conditions of the ideal state disjointly, one after another, however, leads to an overall worse outcome. This is known as the problem of the "second best". To understand this problem of the "second best", we can think back at the "baking recipe"-metaphor and imagine policymaking crudely as a form of cooking or baking. One might think that if one lacks one or more ingredients of the baking recipe, one would still be better-off and at least approximate the desired cake, if one were to bake it without those missing ingredients. Hence, one might—mistakenly—assume that every step toward the fulfillment of a particular desired state (the cake) is an improvement of the current state. Or, put differently: If the ideal state would be comprised of a list of conditions that had to be fulfilled, for example, the absence of executive or in general political power and the compliance of its members with some principles of justice, then one might be tempted to think that revoking executive and political power would bring us instantaneously closer to the ideal state despite the continuous lack of compliance. At least, one of the two ideal conditions would be fulfilled. However, these conditions are in a way interdependent so that trying to fulfill one of them without the other, would turn out to be rather disastrous: Revoking executive power without widespread compliance is like baking a cake without flour, eggs and butter: Better to bake no cake at all. Rather than bringing us closer to justice, it would bring us further away from it. Hence, Estlund suggests that "if we will not comply with a requirement, say, to Build and Comply with certain institutions, then (owing to further facts, such as the consequences of building dysfunctional institutions) it is probably not the case that we ought to build them. The truth about justice could conceivably be without practical implications in this way" (Estlund, 2020, p. 305).

I think the realities of political decision-making are richer than these examples suggest and that utopianism itself—and in particular technological utopianism—suggest ways to sidestep or fix some of the conditions that are allegedly interdependent without the repercussions that some examples of the approximation fallacy suggest. Perhaps executive power is indeed needed without widespread compliance, but would we have to get rid of it from one day to another? Since Estlund accepts some form of the problem of the second best, he wonders whether utopian theorizing—the justification of principles of ideal justice—might be seen as valuable in any other way.

In a first step, he analogizes ideal theories of justice to mathematical theories. Since much mathematical knowledge is practically irrelevant too, but usually considered as intrinsically valuable, so could we consider that an advancement of our knowledge of justice in the form of ideal theories of justice have such positive property. Hence, this route would provide an alternative to my own arguments for the value of utopianism. Its value could be of an intrinsic kind. He writes:

> If you think math can have nonpractical value then you cannot deny a piece of political theory value on the bare ground that it has no practical value. My aim in this section is only to argue that much math that is thought by many (including most of my readers, I suspect) to be important and valuable intellectual work has little or no practical value. In that case, either it has no value after all, or some intellectual work with no practical value can nevertheless be of great value. (Estlund, 2020, p. 310)

However, could that insight carry my argument about worthwhile engagement? Estlund himself seems doubtful. Even if we agreed with the analogy, it remains unclear whether the intrinsic value carries any weight to prompt action, which the thesis of WE requires. To underscore the gap between intrinsic and practical value, Ulla Wessels cites affirmatively a case originally presented by Jonathan Glover (Wessels, 2011, p. 26): Imagine you sit across a man in the train, who starts leaning out of the window while you pass through a tunnel, coloring the wall with a spray can. He insists that it will look beautiful and asks you to join him. You might agree with him that it will indeed look beautiful to colorize the tunnel, more beautiful than without the color. As nobody will ever see it, you might consider its beauty intrinsically valuable. Would one be inclined to join him? Perhaps out of sympathy or politeness, but nothing else. In many cases, thus, intrinsic value seems incapable of inciting action. It seems, therefore, unclear whether engagement can be supported by an argument about intrinsic value. Estlund does not settle with the matter here. He goes on to suggest that it is commonly assumed that having "an informed concern about social justice and injustice is a significant aspect of a morally good person, apart from any value the concern or the understanding might have for use in producing good action or anything else of value" (p. 317). Here, the picture of a moral person is invoked to suggest that we characterize such a person as someone, who seeks a better understanding of justice. That seeking of understanding might entail engagement with various types of theories of justice including ideal theories of justice.

The argument is appealing, yet problematic: First, it carries an air of paradox or question begging. While I claimed before that technological utopias advance our understanding of justice and suggested, hence, that engagement with them is warranted, Estlund seems to turn this view on his head. He suggests that engagement with the idea or concept of justice is naturally sought for by virtuous people. In their quest toward the "right" understanding of justice, they will naturally direct their interest towards ideal theories. But, is not the central question what the right theory of justice is, realistic or idealistic, that underlies the concern of understanding? Radical realists who claim—as we will see in Chapter 4 in more detail—that some forms of ideal theorizing are not concerned at all with justice, but rather engage in thought-experimenting, might simply recommend the virtuous person to search for a better understanding of justice by turning to realist theories.

But, secondly, we must acknowledge, as mentioned before, the vast number of utopias that are not concerned with the justification of ideal *principles of justice* (e.g. Ursula Le Guin, Kim Stanley Robinson, N. K. Jemisin, etc.), for whom this argument is outside of reach. Estlund suggests that Cohen's and Rawls' accounts of justice are the prime examples of ideal theories of justice, who *advocate principles of justice* and in Rawls' case—also justify the particular design of institutions. This narrowing of theorizing about justice, puts us into the following trilemma: We can either (rather forcefully) (a) suggest that other (traditional and modern utopias) are actually, subliminally, also concerned with establishing principles of justice, (b) we could suggest that they improve our understanding of justice in other ways than justifying ideal principles of justice, or, lastly, (c) we could accept that many utopias—including technological utopias—as they fail to justify principles of justice, fall outside the scope of Estlund's arguments and, therefore, require us to establish other reasons for their value.

As the first option seems rather far-fetched, I appeal to a mix of (b) and (c): In my view, utopias can do more than advancing principle of justice and this is valuable for several reasons that Estlund misses to appreciate. Utopias advance a more general idea of justice that is not concerned with principles, but rather with the background assumptions of theories of justice (their scope) and the location of justice—the various nexuses to intervene to establish a just society. They challenge background assumptions such as the idea that human nature—which is often considered to be only to a limited degree altruistic—is immutable and it also challenges the very idea that in order to establish a radically different and better way of living, one would have to "mold" people in ways that undermine their liberty and their human rights. I will underscore this as a particular strength of the outlook of technological utopianism, where technologies are regarded as the means to sidestep the motivational problem that is usually seen as a central problem of utopianism, which will be discussed later on (Chapter 5) (Nagel, 1991). This critiquing of notions such as these (e.g. the immutability of human nature and inevitability of motivational constrains, when pursuing socio-political change), that return as background assumptions in theories of justice, fulfills a cognitive and dispositional function: It advances our understanding and thinking about justice, and such advancement can also entice various forms of social activities (e.g. forming communities, establishing alliances, discussing and developing ideas further, engaging different with the world) that are neither violent nor at all directly engaged with governance.

Scrutinizing the location and scope of justice—which is what technological utopias can provide—can lead to a better understanding of justice. Such insight is cognitively stimulating and practically relevant—both dimensions are inseparably intertwined. This is why we should engage with them. In this sense, we must understand practical relevance as being more than giving concrete guidelines for practical socio-political change or initiating such change. The functions of utopia, in my view, can be roughly distinguished into a cognitive and dispositional dimension, whereby the dispositional dimension refers to utopias' propensity to motivate or incline toward or desire change. The distinction is illustrative or orienting. I do not want to insinuate that they are of a categorically different type. In reality, they might never be fully separable: The disposition to initiate change can be as much a consequence of a creative input, a critical stimulant, that helps a person to look at the world differently and recognize a deficit in our way of being that she has not seen before, as much as—in turn—the creative input can be a product for her sudden desire for change. Neither dimension shall claim a prerogative or deserves to be valued more than the other (Fig. 2.1).

We have already found an advocate of the critical function in Krishan Kumar, who suggests that More's *Utopia* introduces an alternative form of social criticism aside from political argument—namely the presentation of an alternative society (Kumar, 1991). He suggests that this is utopias' subversive function. The presentation of an image of a different way of living and being works like looking into a slightly distorted

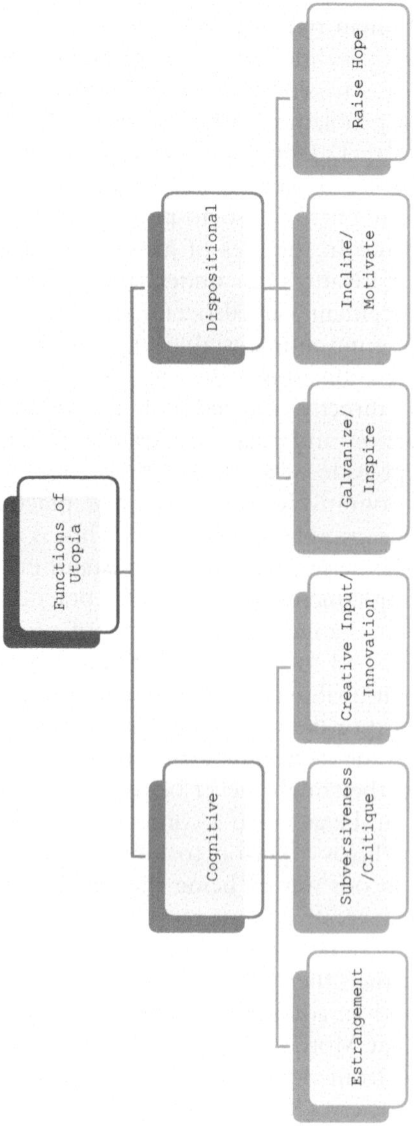

Fig. 2.1 Various functions of utopianism

mirror: We will start wondering, when seeing ourselves on the other side, whether this is how we really look like, or rather would want and should aspire to look like. Both, dissatisfaction or satisfaction could be the product of such confrontation. Dissatisfaction with the present, if we realize that the distorted image in the mirror is what we aspire to be, but are still far away from.[2] Satisfaction, if what we see, pleases us and we can gladly exhale and accept who we currently are and consider how to preserve it (this is the result, dystopias aim to provoke). This transformative potential is clearly not exclusively directed at those who have or want to assert political power narrowly conceived. The process initiated by such confrontation with the imaginary has been—following Frederic Jameson—widely regarded as a process of estrangement (Jameson, 1982, 2005). We might realize when confronted with the image of the alternative world, how things that we have never doubted or put into question are mutable. We are in a sense ideologically imprisoned by such unquestioned assumptions about the immutability of some of our surroundings (Jameson, 2004). When the spell of ideology is broken, we are freed to see the world with different eyes, to recognize its mutability. It is not the same world anymore. Lucy Sargisson has further delineated the idea that is entailed in Jameson's notion that utopianism breaks the spell of perceiving the world as being without alternative by estranging us from the present:

> Estrangement is a generic convention of science fiction utopias which is increasingly important and has purchase beyond literary texts. It is an integral part of utopianism. It permits utopias to function critically and occurs in both the structure and the content of utopian visions. The concept of estrangement combines a number of cognate ideas, including distance, excess and difference. (Sargisson, 2012, p. 18)

Mathias Thaler has recently developed an impressive expansion of utopian scholarship and applied his theoretical insights to climate change visions (Thaler, 2022). Here, I cannot do justice to the breath of his theorizing. Still, I want to present some major lines of reasoning that support my own views and underscore his original contribution to the field. In his earlier works, he emphasizes the cognitive, critical function

[2] The transitioning between the alternative world and reality is in itself part of the storyline of some utopias (e.g. *Roadside Picnic* by Arkadi und Boris Strugazki).

of utopianism. Thaler suggests that "utopian fiction [...] can be useful for comprehending 'what is' (thus helping us to soberly understand the world we currently inhabit) and for meditating on 'what might be' (thus helping us to nurture the hope for a better future). It is in the interplay between these two modes of critical reflection that [utopias'] value for political theory lies" (Thaler, 2018, p. 671). He suggests that looking into the "what might be"—the alternative world of utopia—conduces to our understanding of our current predicament. Rather than being escapist, becoming aloof and permanently distanced from present concerns, the utopian recipient actually utilizes insights into the other to look differently at our present world. Though, he does warn later that succumbing to the utopian mode of engaging predominantly in the "what-if"-mode of theorizing, one might start to dissociate oneself from present concerns (Thaler, 2022). We can assume that this danger might be one that he most strongly assumes lurking in the utopia of virtual reality. With this in mind, he pleas for an estrangement "for the world", rather than "from the world". He suggests that "[by] establishing critical distance from what is conventionally taken for granted, utopias teach us to perceive reality from surprising and illuminating angles. In short, they prompt us to see the world anew. This process has the effect of disrupting habitual patterns of lived experience that hold sway over our normal modes of existence" (2022, p. 43). Here, Thaler strongly intertwines the cognitive function that we have earlier distinguished conceptually from the dispositional function, assuming that the insight into "seeing the world anew" also inclines toward engaging with her differently. This interdependence can be affirmatively viewed. Furthermore, he underscores that utopia brings novelty into existing discussions about how to live and be better: This creative function that introduces new ideas and concepts into debates about justice and into political theory, that I have emphasized before, is often underestimated. Many a political ideal—even many of those that are now seen as belonging to the standard repertoire of realistic political ambition—such as the idea of equality before the law—have at some point been introduced via fancy, utopian speculation. Thaler largely adopts Ruth Levitas' terminology of utopianism as a "desire for a better way of living and being" and combines it with Miguel Abensour's notion that utopia educates this desire regarding where toward to direct itself. In his 2022 book *Utopian Visions for a Climate-Changed World*, Thaler distinguishes three general plotlines of utopian speculation, which he identifies among the various climate change visions, each containing its own benefits and risks, and each accentuates one of the previously distinguished functions:

> The What-if plot line aims to undo the habitual patterns of experiencing the world, but it is susceptible to the charge of indeterminacy - what ought to follow concretely from the strange scenario conjured by these utopias is not always evident. The If-Only plot line, by contrast, seeks to mobilize people into action, yet its pedagogy of desire might come dangerously close to wishful thinking. Finally, the if-this-goes-on plot line intends to warn an audience about imminent risks, its depiction of a dire world to come can veer into fear mongering that incapacitates, rather than motivates, people. (Thaler, 2022, p. 49)

In the increasingly convoluted space of contemporary utopian theorizing (Chrostowska, 2021, p. 5), these distinctions serve as an important instrument for further analysis and understanding. It is unfortunately beyond the scope of the present essay to investigate these distinctions' application to technological utopias of the kind under scrutiny here. Thaler suggests that the ecomodernist project, which has strongest resemblances with the technological utopias discussed here, concurs mostly with the "if-only"-plot line, which is focused on galvanizing change, "charging their proposals with the emotional energy of a positive outlook on the future" (p. 169). *Prima facie*, this judgment seems also applicable to Transhumanism. Echoing the function of subversiveness and critique that other utopian scholars affirm, Richard Saage argues:

> The goals projected [in utopia] are distinguished by a precise critique of existing institutions and sociopolitical relations that stand in contrast to an alternative articulated in secular categories. Without the antithetical confrontation of utopian ideal with what needs to be criticized, the classical utopias would lose their identity. (Saage, 2016, p. 63)

We shall later see that Saage believes that contemporary technological utopias implicitly or explicitly endorse the general socio-political framework (liberal capitalism) and the individualization tendencies of our society. Hence, they have lost the antithetical character that features so prominently in traditional utopias. With this general sentiment, Saage echoes Ernst Bloch, who suggests that utopianism after Marx "has turned into reactionary or superfluous playful forms. These do not lack a seductive quality of course, and are at least useful for diversion, but this is precisely why they have become mere ideologies of the existent [...]" (quoted in: Chrostowska, 2021, p. 4). Though, as we talked earlier about the playful irony already present in More's *Utopia*, one might doubt how widespread this feature is within the classical utopian tradition. In any case, the argument suggests on the one hand that such judgment appertains to technological utopias. While that might be true for some, it seems less plausible for others. Furthermore, Saage assumes that the promotion of liberal capitalist values cannot be understood as a form of utopian speculation. I will suggest that there might be good reasons—given the imperfections and deficiencies of current liberal societies and capitalist markets—that even a form of hyper liberalism might deserve the utopian seal. I will, however, also outline why the supposedly needed antithesis between individualist vs. collectivist perspectives on society that Saage elucidates—while theoretically tenable and perhaps interesting for the organization of historical material—is substantially void. Not only can changes to the social group as a whole be deeply affected by changes to many individuals. Recognizing this, only requires a clearer distinction between the individual as the primary change agent from the individual as the object of change, which can be impacted by adjustments to society and the interaction between individuals. Additionally, clearly understood, technological utopias promote technology as a third pillar aside from individual behavior and socio-political institutions to form the potential locus of a just society. Through this medium, technological utopianism rather exceeds and continues to blur the boundary between individual obligations and socio-political institutions—it is exactly the envisioning of new socio-technical constellations that undergirds the positive value of technological utopianism. Furthermore, they challenge the scope of justice in that they press us to consider as mutable aspects of the world that inform theories of justice and are usually seen as immutable.

Summarizing, we have seen that practical relevance can be narrowly conceived as determining immediate political decision-making and action. Under the impression of Fascist and Soviet communist revolutions, Popper thinks that utopians necessarily aim for practical relevance in such narrow sense and that such must necessarily lead to violence. We have seen that practical relevance can be more broadly understood. There are various functions that intertwine both dispositional and cognitive elements that utopias fulfill. As I shall later suggest that utopias' benefits can offset some of the dangers of (technological) utopianism, including their elitism, we must keep those positive aspects of the practical value of utopianism in mind. But first, there are two more objections under the wider umbrella of fatal perfectionism that deserve close inspection.

2.3 Lock-In, Boredom and Stagnation by Boredom and Stagnation

The political theorist Judith Shklar once wrote that "utopia, the moralists artifact, is of necessity a changeless, harmonious whole" (quoted in: Sargent, 2010, p. 93). This last critique of utopianism as stagnant and boring relates to a more traditional concern (Osborne, 2006). Attention to this criticism has recently grown in connection with comprehensive technological utopias such as Transhumanism that promise—as we have seen—otherworldly degrees of flourishing and well-being—a "solved world" in other words (Bostrom, 2024).

An intriguing discussion of this dimension of the problem of perfectionism has been developed in *Automation and Utopia: Human Flourishing in a World without Work* by John Danaher. He suggests that the possibility of lock-ins and boredom lurk as some of the biggest dangers of striving toward technological utopia (Danaher, 2019). Danaher's treatment of technological utopianism is one of the most comprehensive of a philosopher of technology to date and deserves careful scrutiny. While the book's arguments about meaning in a world without work have been widely discussed, the views on utopianism have only recently been critically picked up (Cea et al., 2023). Danaher is by no means an anti-utopian in the same vein as Popper. He underscores the need to concoct rich, futuristic imaginaries and to scrutinize their upsides and downsides, to prevent us from heading down undesirable pathways (Danaher, 2019, p. 85). In some sense, his endeavor is, therefore, more practically oriented than my own (as outlined in the previous section) and perhaps closer to what Nordmann criticizes as a form of "speculative ethics" (see Chapter 3).[3] He cherishes the many possibilities of currently emerging

[3] Danaher's account is rich and there are more insights that would deserve to but cannot be extensively discussed here. He suggests, for instance, that thinking about utopianism concerns predicting and analyzing which technological futures are possible and how we can evaluate and choose for the best: "The job of the futurist is the job of figuring out which pathway we are going to take through the landscape of possible future worlds. Which possible world is more likely? More technically feasible? And what will that world look like in its totality? Answering these questions is not an easy task. Futurists often fall into the trap of narrow thinking. It is relatively easy for them to describe a possible future

technologies of automation that can (and will most likely to some degree) rid us of forms of work, but also change the world in other ways. They might undermine our inclinations to live meaningful lives and lead us into boredom and stagnation. Danaher discusses whether such prospects should be seen as rather dystopian than utopian.

Danaher's own definition of utopianism is that of an ideal society (p. 82). He introduces in more detail what he considers to be an ideal society via a definition of Christopher Yorke as "[a]ny prospectively achievable scheme of radical social-political improvement which would, if installed, leave every affected party better off and none worse while respecting the rights of all" (p. 84). While the demand for "socio-political improvement" resonates clearly with Levitas' suggestion that utopia expresses a "better way of living and being", the insinuation that such change must be "radical" is not immediately plausible: The realistic utopias of Erik Olin Wright are not of that kind (Wright, 2012). They envision changes that have either been already partially implemented on some smaller scale or can already be easily tested on a smaller scale (in the Netherlands companies have started experimenting with a 4-day work week [while maintaining full salary]) and tested the Universal Basic Income in small-scale experiments (Ghatak & Maniquet, 2019).

Of course, the account raises a number of other questions: Should we really accept that *everyone* needs to be better-off—even those that are currently seen as being already too well off (at least in financial terms)? This seems an excessive demand. Very few traditional utopias might concur that their vision of a better way of living and being, is a better way of living and being for *everyone*. That seems to be a more unique promise of recent, twentieth-century technological utopias, which speculate about the possibility of massive degrees of abundance. Further, do the members of such an ideal society have rights that need to be respected? Which rights

if they just imagine one or two changes to the list of propositions that describe the present reality. But the likelihood is that any one change, particularly if it concerns something major like the development of superintelligent AI, will cause hundreds or thousands of other, difficult to anticipate, small changes to that possible world. Can you imagine all of these (or at least a significant chunk of them) in rich detail? This is something that the best futurists must do. It is also something I must do throughout the remainder of this book. I must imagine and evaluate the different possible futures toward which we could be heading, in rich and plausible detail" (p. 85). In my view, this conflates the job of the utopian with that of the forecaster. It would be worthwhile to consider his recent take on axiological futurism more extensively than present space permits (Danaher, 2021).

are we talking about here and what about clashes between rights (the right to not being denunciated and the right to express oneself freely—how can both be protected?). We will later see (Chapter 4) that political philosophers such as Cohen, who devises his image of an egalitarian society based on ethos and community, see little need for rights-conferring socio-political institutions in utopia. The clause of "achievability" seems in line with my previous injunction that utopianism is not by definition so irrational as to demand what is literally impossible. So, while the definition, therefore, strikes some chords and seems plausible by and large, it remains undefended in Danaher's account. Is it historically informed—what about the counterexamples just mentioned that present non-radical improvements? Such are the definitory concerns that we earlier tried to escape, and which Danaher has sidestepped in his approach. Let us turn to the problem of perfectionism as lock-in and the fallout of stagnation.

Danaher considers different utopias based on the premise that automation will massively reduce the demand for human labor. In the first half of his book, he discusses both the likelihood and desirability of this development. He considers this development as by and large desirable, as he considers work to be structurally bad and believes that life without work could retain meaning. However, as he also sees the risks that emerge in the transition toward a world without work—such as neglect of those who have lost their jobs—before other meaning giving conditions are established, he looks for other opportunities to smoothen both the transition toward a world without work and also to minimize the risks that other problems aside from perceived lack of meaning are mitigated. He suggests that "we should embrace the idea of a post-work future, albeit with some degree of caution. Such a future could be utopian if we get it right" (2019, p. 83). The reference to the conditionality of what type of change is needed, takes him to discuss under which circumstances a world without work would be desirable. How would a post-work utopia look like? Hence, he discusses utopianism in a way that I am defending here as worthwhile engagement. He understands perfectionism as a sort of stagnation—being locked-in into a socio-political reality that perhaps appears pleasurable at first but is entirely devoid of progress. His primary example is based on the movie *Wall-E*. In this scenario, humans live lives of plenty. They have left earth to live on spacecrafts, where they are continuously fed by robots and their physical and mental pleasures are fulfilled at all times. They hang around, listen to music, watch television, get obese and seemingly decline mentally. Danaher suggests that they have reached a state of apathy, unbeknownst to them, which they cannot escape anymore. They are locked-in in a state of conveniences and comforts that technologies supply. They have become so accustomed to this lifestyle that they have lost all urge for change and betterment. Being locked-in in such a state, therefore, means that we have established a way of living that is far from perfect, but rigid or almost irreversible, because of the complacency, which it has caused among its inhabitants. Danaher conclusively argues that "[…] stability can also have a dark side, one that is antithetical to utopian aspirations. Without some sense of unease with the present reality, and without some goals and aspirations for improvement, it seems like we would have little reason to get out of bed in the morning. If a utopia was truly a stable end point of social development, then it would be pretty boring" (2019, p. 92).

To underscore an important point for the evaluation of such scenario, I shall contrast this case—call Danaher's the "loss of urgence"-scenario—with another paradigmatic dystopian scenario that portrays a state of lock-in in another, but similar sense: Huxley's *Brave New World*: The socio-political institutions in *Brave New World* are profoundly coercive, operating through a combination of psychological manipulation, social engineering and chemical control. Unlike more overtly oppressive regimes, the World State in *Brave New World* achieves totalitarian control by shaping desires and beliefs from birth, making the population complicit in their own subjugation. This subtle yet pervasive coercion ensures the stability of the society, but at the cost of individual freedom and true happiness. This is a paradigmatic case of a closed society, whose societal institutions are designed to control their individual citizens, to regulate their use of language, observe and control their everyday behavior, sexual and social life and their thinking and, thereby, undermine all individuality, but also all forms of creativity and, it seems, intellectual progress. John, a figure who comes from a small community outside the World State, when coming across this way of life "is horrified by the shallow, hedonistic passivity of its citizens. Lacking art, religion and any sort of genuine passion or curiosity, this stagnant society has, John says, paid 'a fairly high price' for its empty happiness" (Ball, 2013). Evaluating the two scenarios side by side, the conclusion suggests the same: Undesirable stagnation, boredom and lack of social development are a byproduct of reaching a state of utopian perfection—with dystopia as a fallout. Yet, the sources of these two lock-in scenarios are different: On the one hand, they come from political suppression and non-violent indoctrination that has been meshed via various dimensions into the fabric of the state that undermine any dissenting tendencies (*Brave New World*). But it can also come, as in *Wall-E*, from reaching a state of abundance and excessive well-being, possibly as a result of voluntary collective choice rather than political oppression.

Juxtaposing these two examples—Danaher's "loss of urgency"-scenario and *Brave New World*—suggests crucially that not all lock-in scenarios are the same and not all of them equally bad or undesirable. Lock-ins in which individuals have found exceptional levels of comfort as in *Wall-E* have apparently led to a loss of the *inclinations* for change. That indicates that their society has become—in Danaher's words—a truly "stable end point of social development", *not* because they cannot be any different, but because they *have lost the desire* for social development and change. They do not *want* to be different. This state—as far as we can tell—might have very well come about *voluntarily* and remains a state in which, presumably, voluntary withdrawal from the established state of pure leisure and laziness *is possible*, albeit unsought for.

Brave New World's state machinery proposes a very different kind of political texture. Although the society in *Brave New World* appears to lack the brutal police state often found in other dystopias (e.g. Orwell's *1984*), it is nonetheless highly effective at preventing dissent. Individuals who do not conform, such as Bernard Marx or Helmholtz Watson, are exiled to isolated islands, where they can no longer influence others. The state also uses subtle forms of surveillance and psychological manipulation to undermine political upheaval from the outset. This scenario does not only mark the *absence of deviant behavior*—of creative acts—it sanctions deviant behavior and punishes it. The badness of this form of lock-in stems from its *undermining of crucial liberties* and, therefore, *the possibility of progress*. Quite possibly, it is also a result of political interventions that have been *imposed on* their citizens rather than having been voluntarily chosen for. To me, therefore, the latter state of stagnation is more obviously undesirable than the former.

Still, the former might be undesirable for a different reason: A life that is devoid of engagement with activities that produce something of value—a life purely of leisure and excess—might be seen as *meaningless*. In the hybrid view of meaning in life, as for instance promoted by Susan Wolf, which is now widely adopted (also e.g. by Danaher himself [2022]), it is argued that both a subjective and an objective dimension need to be met (Wolf, 2010): The subjective dimension refers to the fulfillment that an individual gains from engagement with certain projects and activities. The objective dimension refers to projects' or activities' value increasing or value conferring capacities. Regarding this dimension, Wolf writes "that the project or activity must possess a value whose source comes from outside of oneself – whose value, in other words, is in part independent of one's own attitude to it" (Wolf, 2010, p. 37). Prime examples for this dimension are striving for justice, doing science or creating works of art. Usually such activities are seen as objectively valuable not only because they satisfy the individual and her intrinsic scientific, artistic or altruistic motivations, but also because they improve the world in various ways. They can bring joy (art), provide valuable socio-political critique for betterment (art), making the world more just, people healthier, more insightful and better-off in general (science).

Obviously, there is plenty of subjective fulfillment in *Wall-E*, or so it seems. There is pleasure and vast amounts of comfort. So, what about objective value? This depends a bit on our interpretation of the objective dimension of Wolf's account of meaning and to which degree we allow our imagination to fill the gaps of the story. According to Wolf, the objective dimension could be fulfilled by projects that embrace "both positive relationships with family and friends", which is a goal unlikely to be ever fulfilled—caring for and maintaining such relationships seems like and infinite endeavor. But, in a more extreme interpretation, which we might concoct—we could consider the world of *Wall-E* as being devoid of the need for any objectively value conferring projects.[4] Imagine that perfect health, absence of hunger and inequality and perfect justice have indeed been completely achieved. So, assume that the story indeed suggests that in this imaginary world—there are no objective values to be achieved anymore, because flourishing is ubiquitous. In such a world then, there can be and need not be engagement with activities that are conducive to foster objective value. Either, perhaps all human life has become meaningless in such a world. Or, in fact, a very different conclusion could be warranted: Namely that the question about meaning in such a world has

[4] One could claim that such outlook is itself paradoxical and somewhat impossible: The arts might be infinitely advanced and objective value conferring. But that is a strong assumption. Obviously, the robot in *Wall-E* is cleaning up an environmentally degraded Earth, perhaps to make it inhabitable again. That in and of itself might seem like an objectively valuable task—restoring the aesthetically pleasing guise of nature. Wolf acknowledges this: "[…] preserving a place of natural beauty all seem intuitively to deserve classification as valuable activities" (2010, p. 37).

become obsolete. In other, more realistic scenarios, it will remain the case that the world can in crucial ways be made better—hence, that meaningful activities will remain possible, if the socio-political order allows for engagement with them.

Conclusively, I do not share the concerns about the dangers of lock-in. If stagnation and boredom are not intrinsically bad—as I have just insinuated—it seems that striving for a utopian perfect state of living and being remains a permissible undertaking, as long as the fancied socio-political institutions would allow for creative leaps and freedom to engage in meaningful projects. We have seen that for many utopian scholars perfectionism is not a necessary condition of the definition of utopia (Sargisson, 2012). Max-More suggests that Transhumanism is a dynamic, procedural utopia that is not interested in envisioning ideal end-state of human development (More, 2013). While these views seem plausible, the previously developed arguments stand on firm ground even without endorsing any such assumptions about particular utopias or utopianism in general.

However, could it be perhaps in some sense *epistemically mistaken* to entertain the belief in the possibility of conceiving such world? This is the last charge against utopianism that I wish to discuss in this chapter, before moving on to criticism more tightly associated with contemporary technological utopias.

2.4 Untenable Epistemological Absolutism

This perfectionist critique has been advanced by Amartya Sen among others. The critique commences by reminding scholars interested in theories of justice that they must take seriously the fallibility of their own viewpoints on intricate problems of justice. He suggests that views on justice are incompatibly pluralistic; each of them can claim with good reasons some degree of plausibility that cannot be reconciled or overcome by way of aspiring toward an ideal by way of rational deliberation. He suggests that "[t]here are *genuinely* plural, and sometimes conflicting, general concerns that bear on our understanding of justice" (p. 57). Sen goes on using the following example to illustrate this. He writes:

> [...] you have to decide which of three children – Anne, Bob and Carla – should get a flute about which they are quarrelling. Anne claims the flute on the ground that she is the only one of the three who knows how to play it (the others do not deny this), and that it would be quite unjust

> to deny the flute to the only one who can actually play it. If that is all you knew, the case for giving the flute to the first child would be strong. In an alternative scenario, it is Bob who speaks up, and defends his case for having the flute by pointing out that he is the only one among the three who is so poor that he has no toys of his own. The flute would give him something to play with (the other two concede that they are richer and well supplied with engaging amenities). If you had heard only Bob and none of the others, the case for giving it to him would be strong. In another alternative scenario, it is Carla who speaks up and points out that she has been working diligently for many months to make the flute with her own labour (the others confirm this), and just when she had finished her work, 'just then', she complains, 'these expropriators came along to try to grab the flute away from me'. If Carla's statement is all you had heard, you might be inclined to give the flute to her in recognition of her understandable claim to something she has made herself. (Sen, 2010, p. 13)

The example contrasts some underlying key values that might determine the way of distributing various goods. Those are roughly speaking; desert (the one who made it), utility (the only one who can use it) and need (the one who has no further toys to play with). Principles of distributive justice based on these various key values have been defended in various shapes and forms. Notable here is that the thought-experiment commences on a few more or less explicit premises that I earlier called background assumptions of justice, which demarcate the scope of justice in my understanding. One of them being that there is *only one flute* to distribute. Hence, we are being asked to make the distributive decision on the assumption of *relative scarcity*. But there are more: We are asked to consider "giving" as some sort of property transaction. In other words, giving one child the flute in the example seems to permanently make impossible that the others can have it.[5] Conferring the object seems to imply that it remains unusable, at least temporarily, for the others. Why would we accept this when conferring property to children? Could we not setup a scheme—with the help of technology, perhaps—that organizes the sharing of the object for equal time intervals?

[5] Another one, is incomplete knowledge: How well can Anne play? Perhaps so well that Bob and Carla voluntarily give up their claims? How much work did Carla put into making it? Would that offset how well Anne can play—how well she would entertain the others with her playing?

While clearly much utopian thinking that is associated with ideal theorizing in political philosophy aspires to determine what would be best, or most just to do in such situations, another gist of much utopian reasoning seems to be to challenge exactly the rigidity of those types of assumptions; scarcity (which we introduced through Rawls' assumption that ideal justice requires favorable historical and natural conditions) and the idea that owning property implies or comes down to (temporar or permanent) access restrictions for other individuals, each of which are widespread. But are they necessary or are we assuming that they are impossible to overcome because we find it unlikely that they will be overcome? I will in Chapters 4 and 5 suggest that such background assumptions of justice constrain the scope of justice; how substantive individuals in a just society can flourish and how many undesirable elements of the world are mutable. In my view, it is a key function of utopian thinking to press political theories to ponder the absence of such background assumptions and fancy larger scopes of justice. That, as I claim, is what many technological utopias do.

On the following page, Sen concludes: "The general point here is that it is not easy to brush aside as foundationless any of the claims based respectively on the pursuit of human fulfilment, or removal of poverty, or entitlement to enjoy the products of one's own labour. The different resolutions all have serious arguments in support of them, and we may not be able to identify, without some arbitrariness, any of the alternative arguments as being the one that must invariably prevail" (p. 14). Sen's argument, thus, charges the utopian of taking the stance of an epistemic absolutism. His view, therefore, holds that perfectionism is misguided in that regarding the most basic and intricate problems of justice—such as exemplified in the case of the distributive justice question regarding the flute—there will always prevail reasonable disagreement about what principle would be best to employ. Anyone—and that seems to be a particular problem in what Sen considers transcendental theories of justice—who suggests that they have by whichever way of reasoning found grounds that either of the aforementioned theories or key values of justice (need, desert, utility) ought to prevail over the other unreasonably "brushes aside" the other options. But what exactly constitutes the badness of such alleged epistemic absolutism?

Sen's argument is developed in opposition to what he considers as "transcendental theories of justice", not necessarily in opposition to what we discuss here as utopianism. Theories of justice aspire to find ideal views on justice. They aspire to determine the right concept of justice—or in other words—what would be perfectly just. There are two ways to understand the presented argument. First, it could be seen as a variation of the claim that utopianism is *unrealistic*: The argument suggests that utopians—idealists regarding principles of justice—believe *wrongfully* that the reasonable disagreement that prevails about the ideal of justice can be resolved. What this reading of the argument would, however, commit to right away, would be to beg the question why we think that these are unresolvable, or the fact of the matter that they are currently unresolvable using as evidence for the impossibility of resolving them in the future. This seems to be the point of contention between realists and idealists.

It seems to me that the idea that we remain necessarily divided about such problems of justice, seems like another background assumption, whose persuasiveness utopians continuously challenge. Promoters of virtual reality, in particular, might claim that their technology allows for a reconciliation of fundamentally opposing ideas about the good life and justice without having to resolve them. The passage might be understood in a different way, suggesting that any proposed solution to the distribution problem would not find widespread support and would have to be implemented against the will of those who disagree. This is where we return to Popper and his assumption that utopianism necessarily becomes coercive and violent. However, it seems unlikely, as Sen makes believe, that is uniquely tied to only this theoretical outlook of transcendental reasoning. Many of us believe certain acts and political decisions to be (absolutely) wrong, and still continue to find shared solutions with friends and relatives, who think fundamentally different.

We must not deny that utopians often flirt with a stance of normative superiority of their outlooks over others. They believe they have indeed some insights into what constitutes—perhaps not a full-fledged just or perfect society—but perhaps a better way of living and being. Barbara Goodwin suggests that it is a matter of normative consistency that one ranks one's theoretical preferred views above others: "Idealism and consistency demand that the utopian should rank his utopia above other conceivable forms of society [...]" (Goodwin 1980, p. 385). However, this just invites us to ask in return, whether Sen's assumed advantageousness of non-transcendental reasoning and social choice theory is not *also* a form of assumed epistemic absolutism. If it is true that normative consistency requires some form of conviction into the superiority of one's views, the real question then becomes whether utopianism succumbs to the other accusation in Sen's argument, namely to "brush aside as foundationless", in Sen's words, competing reasons and views. That is what he might underscore as the major differences between his preference for social choice theory, and the utopians preference for transcendental or other forms of theorizing. How would such "brushing aside" exactly express itself and what would it amount to?

As indicated before, I believe the problem of maintaining normative consistency while pursuing the goal of resolving normative conflict is not only widespread, but also that they appertain to the more realistic ways of reasoning about what to do—and a stance of epistemic absolutism is often rather naturally adopted. Consider some parents who disagree about which school would be best for their kid. Each without contradiction might firmly believe that their preferred choice, for which they might have provided extensive reasons, is the better one, or in other words regarding the context: The best school to send the kid to. This—while being a very realistic concern, a very practical question so to speak—does not concern an abstract discussion about the nature of justice; yet, the parents clearly take a stance of normative superiority. Each believes they are right. However, there is no contradiction in each of them also assuming that their partners have genuine and genuinely *good reasons* for their own preferred choice. Yet, in their views, their own reasons are more convincing. Eventually, they must figure out how to decide, either by ameliorating their views through advanced procedures and improved communication (making the reasons transparent by putting them on a list) or by a decision-making procedure that gives each the same chances (throwing dice). Both can in fact without pain of inconsistency accept "normative superiority" of their own views, without endorsing that the other's position can be brushed aside as "foundationless". This shows that the problem Sen describes is far from being unique to the context of utopian reasoning. It is not so much a result of transcendental argument or idealization as it might concern the very nature of normative conflicts. And oftentimes, such conflicts regard the very question, whether one of the opposing views is too idealistic or realistic. What this also suggests, is that alleging a "normative superiority" does not entail "brushing aside as foundationless" other perspectives. In such everyday situations, while holding on to one's views, one can readily acknowledge the overarching need for a shared solution and that such requires establishing a third way or yielding for the purpose of practical advance.

Sen suggests that rational choice theory provides an alternative to resolve problems of practical political decision-making. Rational choice theory is devoid of epistemological absolutism that characterizes utopian thinking, insofar as it does not aspire to determine a perfect solution, only the best choice among several available, realistic alternatives. Rather than devising what would be the best outcome (e.g. under circumstances that are unrealistic), it provides comparative assessments commencing from self-evident axioms. Sen suggests that such rational choice theory has various advantages over forms of transcendental reasoning, viz. utopian thinking. Again, it is important to underscore that Sen's attack is directed at the Rawlsian way of approaching a theory of justice—namely by justifying the right and ideal principles of justice by way of transcendental reasoning. Whether the attack finds a suitable target in the other realms of utopian thinking that is vastly different in outlook, style and aspiration is rather questionable.

David Estlund suggests that rational choice theory can provide an important way of approaching decision-making situations, where choices must be made conservatively, between already available and realistic alternatives. However, he continues to argue that does not mean theorizing about justice must remain in the same sense conservative: "[…] even if choice ought to be conservative, it doesn't follow that theory ought to be conservative" (Estlund, 2020, p. 268). I agree: Utopias of various kinds can provide creative input toward the rational choice models that are employed in political theory and economics. Taking the utopian literature into consideration could help us understand, why alleged self-evident assumptions (concepts of rationality that underlie the rational choice models) are less self-evident than they seem and what other choices we have that we might have missed out considering with a certain model.

The previous survey of criticism of utopianism in general—which equally challenges technological utopianism as a subspecies of this genre—has made the first steps in strengthening the thesis of WE. Via a critical reading of the anti-utopian charges, we did not only find reasons to reject their criticism, but also introduced some constructive points regarding the practical value of utopianism. Hence, the previous approach follows in the footsteps of earlier anti-anti-utopian endeavors (Jameson, 2005). We have seen that utopianism—if we do not confine it to a project of establishing principles of justice—can advance the idea of justice more generally. Rather than having to search for arguments for its intrinsic value, we recognize—and will continue to clarify further later on—how estrangement and subversiveness unravel in the realm of theorizing about justice.

References

Ball, P. (2013). In Retrospect: Brave New World. *Nature, 503*(7476), 338–339. https://doi.org/10.1038/503338a

Bostrom, N. (2024). *Deep Utopia: Life and Meaning in a Solved World*. Ideapress Publishing.

Cea, I., Seeger, A. L., & Wachter, T. (2023). Technological Unemployment and Meaning in Life, a Buen Vivir Critique of the Virtual Utopia. *Humana Mente, 16*(44).

Chrostowska, S. D. (2021). *Utopia in the Age of Survival—Between Myth and Politics*. Stanford University Press.

Danaher, J. (2019). *Automation and Utopia Human Flourishing in a World without Work*. Harvard University Press. www.jstor.org/stable/j.ctvn5txpc

Danaher, J. (2022). Virtual Reality and the Meaning of Life. In I. Landau (Ed.), *The Oxford Handbook of Meaning in Life* (pp. 508–524). Oxford University Press.

Estlund, D. (2020). *Utopophobia—On the Limits (If Any) of Political Philosophy*. Princeton University Press.

Ghatak, M., & Maniquet, F. (2019). Universal Basic Income: Some Theoretical Aspects. *Annual Review of Economics, 11*, 895–928. https://doi.org/10.1146/annurev-economics-080218-030220

Goodwin, B. (1980). Utopia Defended Against the Liberals. *Political Studies, 28*(3), 384–400. https://doi.org/10.1111/j.1467-9248.1980.tb00476.x

Goodwin, B. (2008). Taxation in Utopia. *Utopian Studies, 19*(2), 313–331. http://www.jstor.org/stable/20719904

Jameson, F. (1982). Progress Versus Utopia; Or, Can We Imagine the Future? *Science Fiction Studies, 9*(2), 147–158. http://www.jstor.org/stable/4239476

Jameson, F. (2004). The Politics of Utopia. *New Left Review, 25*.

Jameson, F. (2005). *Archeologies of the Future: The Desire Called Utopia and Other Science Fictions*. Verso.

Kumar, K. (1991). *Utopianism*. Open University Press.

Kurzweil, R. (1999). *The Age of Spiritual Machines: When Computers Exceed Human Intelligence*. Penguin.

Mao, D. (2020). *Inventions of Nemesis: Utopia, Indignation, and Justice*. Princeton University Press.

More, M. (2013). The Philosophy of Transhumanism. In M. More & N. Vita-More (Eds.), *The Transhumanist Reader* (pp. 3–17). Wiley-Blackwell.

Murphy, L. B. (2000). *Moral Demands in Nonideal Theory*. Oxford University Press.
Nagel, T. (1991). *Equality and Partiality*. Oxford University Press.
Negroponte, N. (1995). *Being Digital*. Knopf.
Osborne, P. (2006). The Dreambird of Experience: Utopia, Possibility, Boredom. *Radical Philosophy, 137*, 36–44.
Popper, K. R. (1986). Utopia and Violence. *World Affairs, 149*(1), 3–9. www.jstor.org/stable/20672078
Popper, K. R. (2012). *The Poverty of Historicism* (2nd ed.). Taylor and Francis.
Rawls, J. (1999). *A Theory of Justice* (Revised ed.). Harvard University Press (Belknap).
Saage, R. (2013). New Man in Utopian and Transhumanist Perspective. *European Journal of Futures Research, 1*(1), 14. https://doi.org/10.1007/s40309-013-0014-5
Saage, R. (2016). Is the Classic Concept of Utopia Ready for the Future? In S. D. Chrostowska & J. D. Ingram (Eds.), *Political Uses of Utopia: New Marxist, Anarchist, and Radical Democratic Perspectives* (pp. 57–79). New York.
Saage, R. (2019). Das Ende der politischen Utopie? In R. Saage (Ed.), *Das Ende der politischen Utopie?* (2nd ed., pp. 13–25). Suhrkamp.
Sargent, L. T. (2010). *Utopianism: A Very Short Introduction*. Oxford University Press.
Sargisson, L. (2012). *Fool's Gold?: Utopianism in the Twenty-First Century*. Palgrave Macmillan. https://doi.org/10.1057/9781137031075
Schütt, H.-P. (2010). Nachwort: Utopie - immer und überall. In U. Arnswald & H.-P. Schütt (Eds.), *Thomas Morus' Utopia und das Genre der Utopie in der Politischen Philosophie* (pp. 305–308). KIT Scientific Publishing.
Sen, A. K. (2010). *The Idea of Justice*. Penguin Books.
Thaler, M. (2018). Hope Abjuring Hope: On the Place of Utopia in Realist Political Theory. *Political Theory, 46*(5), 671–697. https://doi.org/10.1177/0090591717740324
Thaler, M. (2022). *No Other Planet: Utopian Visions for a Climate-Changed World*. Cambridge University Press. https://doi.org/10.1017/9781009030250
Valentini, L. (2012). Ideal vs. Non-ideal Theory: A Conceptual Map. *Philosophy Compass, 7*(9), 654–664.
Wessels, U. (2011). *Das Gute: Wohlfahrt, hedonisches Glück und die Erfüllung von Wünschen*. Vittorio Klostermann.
Wolf, S. (2010). *Meaning in Life and Why It Matters*. Princeton University Press.
Wright, E. O. (2012). Transforming Capitalism Through Real Utopias. *American Sociological Review, 78*(1), 1–25. https://doi.org/10.1177/0003122412468882

Žižek, S. (2008). The Violence of the Liberal Utopia. *Distinktion: Journal of Social Theory*, *9*(2), 9–25. https://doi.org/10.1080/1600910X.2008.9672962

CHAPTER 3

Technological Anti-anti-utopianism

Abstract Critics claim that technological utopias are flawed in particular ways: Their proponents are elitist and ruthless. While some of this criticism can be equally leveled at more realistic forms of socio-political engagement, there are concerns about the roots of technological utopias that must be taken seriously. Still, the overarching benefits of technological utopias might well outbalance those concerns. We will unpack the idea of technological utopias' social obliviousness, which will pave the way toward our discussion of utopia and the scope and location of justice.

Keywords Elitism · Social obliviousness · Individualist vs. collectivist utopias · Prediction

In the previous chapter, I have positioned myself *vis-à-vis* the existing literature on utopianism, although I cannot even remotely claim having addressed it comprehensively (which is an impossibility given the scope of each of the fields that show an interest in (technological) utopias). I have discussed several criticisms (Popper, Danaher and Sen) and objections against *utopianism in general* and sometimes alluded how the criticism could relate to technological utopianism. Throughout the discussion, we have advanced through an anti-anti-utopian dialectic a more positive outlook on utopianism, which requires further support.

M. Sand, *Technological Utopianism and the Idea of Justice*,
https://doi.org/10.1007/978-3-031-75945-1_3

In this pursuit, I will focus in the present chapter on criticism that is decidedly directed at *technological utopias* and *engagement with* technological utopianism. This will help to pave the way to my constructive view that will be expanded in Chapter 4. At times, this requires reading some critics of technological utopianism perhaps stronger than they themselves intended to be read. Some of the presented commentators have launched their attacks with a particular technological utopia in mind (e.g. Morozov's attack on "cyber-utopians"); but their points are often of a more general nature, since they apply to all narratives that have similar features or makeups. It is perhaps helpful to restate the thesis that ought to be developed here once more: Despite the many dangers and pitfalls of technological utopianism that will be discussed in the following—some of which, as I will concede, are real—the emphasis of those dangers has overwhelmingly produced a risk-focused discussion. This has led to the omission of discussing the *benefits* of technological utopianism. For any discussion about the value of something, such as utopias, if we consider only their risks, we determine—if anything—their value only *pro tanto*. We must also consider their benefits, otherwise the overall evaluation has significant shortcomings and blind spots. In pursuit of an *all things considered* judgment, I will show that—aside from the very relevant and illuminating ideas that have been brought forward in the Utopian Studies literature, which we have already surveyed before—there are indeed additional reasons to engage with technological utopias, namely their potential to envision new socio-technical constellations that advance our understanding of the scope of justice and e.g. their conviction in the possibility of bypassing the motivational problems that undermine the realization of a better society, which is an example for a background assumption of justice theories. One could also understand my argument not so much as an addition to those aforementioned arguments but as a delineation of how they unravel in the context of technological utopianism. In previous chapters, I have suggested that my view is not a general vindication of technological utopianism, insofar as many technological utopias clearly lack the features from which the positive value according to my theory emanates.

It is worth being a bit more precise about what I mean here with engagement—earlier introduced as a part of WE. Engagement means the attentive reception, careful and charitable analysis of those narratives, inspection of whether and how they identify shortcomings in our current socio-political and technological situation that we might have misunderstood or overlooked, and what constructive suggestions they make to alleviate those shortcomings. This rough characterization should suffice. It will be difficult to specify what that means in practice; whether, for example, one ought to read the Transhumanist manifesto rather once or twice. The general idea is that people with the relevant backgrounds or an interest in the future of human society (this together already includes very many people, including political scientists, ethicist and Technology Assessment [TA]) have reasons to take these narratives serious and try to understand and consider what one can learn from them and grasp their major characteristics.

The type of engagement that I have in mind is *deep* insofar as it differs from a mere sociological, descriptive analysis. Vincent Mosco, rather than using the concept of a utopia, suggests that cyberspace is a myth in the way in which Roland Barthes understood the concept "myth". Mosco suggests that we should have an anthropological interest in those narratives and that it would not be entirely pointless, but very limited, if we were to discuss whether those myths are true: "[…] myth are neither true nor false, but living or dead. […] To understand a myth involves more than proving it to be false. It means figuring out why the myth exist, why it is so important to people, what it means, and what it tells us about people's hopes and dreams" (Mosco, 2005, p. 29). I agree that all these aspects and dimensions constitute important research interests in themselves—not only regarding the evolution and dynamics of myths, but, of course, also regarding our present topic of utopias. I also agree that "true" and "false" are evaluative categories that are not squarely applicable to either myths or utopias. But, these are not the only normative concepts that might be suitably applicable here and such normative questions are as important as sociological, descriptive ones. They naturally arise before and after an advanced understanding of the social embeddedness of a certain vision has been achieved. Unsatisfied with a mere descriptive understanding, this normative inclination presses us to ask whether there is anything we can learn *from them* and what that is and not just whether there is anything *about them*. In the literature, aside from the arguments that attack utopianism in general, we find three reasons to *not* engage with technological utopias. The following titles provide *in nuce* sketches of those critical theses, which will subsequently be presented in more detail and critically examined.

3.1 Technological Utopians Are Ruthless and Elitist

This view has been prosed by, among others, Langdon Winner. He underscores on various occasions that technological utopians are guided by ulterior motives such as financial gain. Others argue that the advocacy of technological utopias has potentially devastating effects: Contra traditional utopians as discussed in the last chapter, technological utopians pursue the realization of their visions, and by doing so, they lead us toward negligent and reckless technological policies. In particular Transhumanists seem eager to realize their vision and push technological development through various positions of power and influence (Hausteiner, 2022, pp. 127 f.; Sand, 2019; Segal, 2007). I shall concede in the following that the elitism that characterizes many aspirations of technological utopians contains risks that shall not be neglected. However, often the alleged dangers of such practical orientation of contemporary technological utopians are overblown and easily outbalanced by the advantages they also bear.

We find some of the aforementioned charges bundled and in an interesting fashion intertwined in a sidenote of S. D. Chrostowska in her *Utopia in the Age of Survival—Between Myth and Politics*, where she explicitly addresses the Transhumanist utopia. She writes:

> The second option [besides 'real utopias'], meanwhile, vests its utopian hope in cutting edge technology; it conceives of transhumanism, eugenics, and life extension as quasi-transcendental emancipation. *It does so not only be deferring the fulfillment of corporal desires here and now*, limited to the somatic makeup of our species, but also by *ignoring emancipation's ethical, political, and economic dimensions*. In its current forms, desire for immortality has been written by the history of domination. It resolves into a desire to escape – rather than eliminate – moral, political, and economic oppression. (Chrostowska, 2021, p. 71; own emphasis)

Much could be said about this passage as it prompts various questions for clarification: Why is the desire of immortality written by a history of domination? The idea seems to follow in the footsteps of a psychoanalytic reading of the wish for longevity by Frederic Jamieson (Jameson, 2005). Still, also other authors have homed in on the alleged corporate agendas of Transhumanists and other technological utopians. Such sentiments have been expressed by Vincent Mosco, who addresses the cyber-utopians of computerization and information, and by Howard Segal, who—with seeming reference to all current technological visionaries—underscores the difference between their motivations and those of traditional utopians:

> The magic wand of computer communication is undeniably seductive. It is also undeniable that much of the allure is manufactured by the very companies that stand to benefit from the sale of computer technology, software, and access to cyberspace. (Mosco, 2005, p. 41)

> Once, forecasting was the province pf people, who regardless of their particular strategies, had a genuine hope of improving the world – or at least a small part thereof. Now it has become an almost purely commercial enterprise intended only to make money. If [...] economists make forecasts, not because they know, but because they're asked, self-proclaimed visionaries seduce a culture that, beneath its surface sophistication, seeks simple reassuring answers to complex unsettling questions. Traditional utopians (and anti-utopians) who cared about the future, no matter how they might have foreseen it, have been replaced by another sort – men and women who care primarily about getting rich and playing off anxieties over the future. (Segal, 2006, p. 69)

More expansive empirical-sociological studies will have to show how widespread corporate interests are underlying technological utopianism and how massively they have influenced the shape and form of those narratives. While there is no evidence that suggests that early Transhumanists, such as Huxley, Haldane or Teilhard de Chardin pursuit anything like corporate interests, it is undeniable that in complex capitalist societies, futuristic visions and predictions are adopted as valuable instruments to constructively manage this complexity. Economic actors utilize those visions to position themselves as a relevant actor in the respective field and to establish alliances. Jens Beckert has extensively studied these dynamics and suggests that

> [...] fictional expectations can help economic actors work in concert in the face of uncertainty: if they share a conviction that the future will develop in a specific way and that other actors will thus behave in foreseeable ways, they may use these expectations to coordinate their decisions. Imaginaries of the future are thus a crucial component of social and economic order. By coordinating action, they also contribute to the dynamics of capitalism, since the correspondence of expectations, or 'frame alignment' [...], anchors decisions for investment and innovation. (Beckert, 2016, p. 11)

But, we have to ask immediately: Is this instrumentalizing role of technological futures and utopias (a) reason enough to attribute ulterior motives to agents who are utilizing it? And, (b) are ulterior motives mutually incompatible with higher ones? Can one not truly believe in one's product as the salvatory tool for a host of societal problems and hope to make money with it? To me, the fervor with which many technological entrepreneurs promote their visions (and in doing so their products) reflects serious convictions in their power. They truly believe in the benefits their technologies will provide and are not—as Segal suggests—in the business of seducing crowds and provide answers that are reassuring and necessarily simple. For utopians—following any meaningful conception of the term—it must be the case that they are not merely interested in the outline of a certain scenario. It seems that if a certain outline of a different way of living and being ought to be considered as a *better way of living and being*—this requires some normative commitment by its proponent. I did not find evidence for dishonesty in the authors that I have studied in the preparation of this essay.[1] Most important, however, is that charging (technological) utopianism specifically with hypocrisy or bad intentions is one-sided. There is no reason to believe that ulterior motives cannot also underlie *more realistic scenarios*. In short, it is unreasonable to assume that the problem that is brought to the fore here undergirds only technological utopians or only utopians in general! If we assume that bad intentions, or selfishness, are problematic when entering discourses and see them as some way of misleading, that is not necessarily connected to utopianism, and it is not necessarily disconnected from realism either. Most likely, we must assume that both proponents of idealistic and realistic visions of a better way of living and being have on occasionally ulterior motives.

We have seen before that traditional utopians have not always been practically interested—interested in realizing their outlines and conceptions (Kumar, 1991). This has changed to some degree in early modernity—in the industrial age—when living conditions have become excessively miserable for many people and scientific and technological progress

[1] "Yet, he [Buckminster Fuller] was certainly different from contemporary 'prophets', who provide not genuine moral critique and who make no serious effort **to alter society for higher purposes**" (Segal 2007, p. 70; own emphasis).

seem to have held the key for real betterment (Saage, 2019).[2] Nowadays, this practical impetus in the more general utopian literature seems to be retreating again (Sargisson, 2012). It is no doubt that among those technological utopias that have been presented in Chapter 1, many are interested in initiating actual and practical changes in contemporary society. They are far from being interested in merely thought-experimenting about possible futures.

In his *Letter to Utopia*, Bostrom suggests to the letter's reader—their former human self: "I can pass you no blueprint for Utopia, no timetable or roadmap. All I can give you is my assurance that there is something here, the potential for a much better life" (Bostrom, 2008, p. 5). Yet, they go on to remind their earlier self that, while no roadmap can be provided, their earlier self can really impact and conduce to utopias' realization: "Whether this surpassing possibility becomes a reality is something you can influence". Others, like Max More, suggest that Transhumanists, while not saying much about "specific goals or measures," are defined by a "commitment to **shaping** fundamentally better futures as defined by values, goals, and general direction [...]" (More, 2013a, p. 28; own emphasis). Michael Hauskeller characterizes Transhumanism, therefore, as a practice-oriented philosophy:

> Transhumanism is without doubt a philosophy with strong utopian tendencies, both in motivation and in outlook [...]. It is a practice-oriented, increasingly influential philosophical-political movement whose proponents and allies frequently and quite openly declare themselves to be motivated by a desire to create a better world or make this world 'a better place'. (Hauskeller, 2014, p. 101)

However, what this "shaping" that More speaks of, and Hauskeller's "practice-orientation" really entail, remains ambiguous. It might either relate to the rather innocuous ways of utopian functions that have been

[2] "Dieser Ablösung von der Raum- durch die Zeitutopie lag die Prämisse zugrunde, dass die Utopie zum künftigen 'Telos' des historischen Prozesses erhoben wird. Die Erweiterung ihres Verwirklichungsmechanismus durch die geschichtsphilosophische Begründung veränderte den Geltungsanspruch der politischen Utopie grundlegend. Sie hörte auf, ein blosses regulatives Prinzip zu sein, das die einzelnen zu grösserer Vollkommenheit anhält. Vielmehr erhebt sie den Anspruch, das in die Zukunft projizierte Ziel auch tatsächlich verwirklichen und eine konkrete politische Transformationsstrategie angeben zu können" (Saage, 2019, p. 18).

outlined before (e.g. initiating reflection that estranges from the present, challenging our ideological imprisonments, creatively advancing discussions about justice and other political concept, etc.). This kind of minimal activity that seems inherent in utopian thinking seems to be shielded from Popperian fears of instigating violence, as has been outlined before.

However, the idea of "shaping better futures" and of "practice-orientation" can also refer to a more proactive approach, where community building and socializing in associations (e.g. Humanity+ which aim at "support[ing] the development of high-impact technology to make beneficial futures attainable.as much as promoting" [https://www.humanityplus.org/]) in addition to promotion, advocacy, lobbying and doing actual research and science become central. In this aspect, then Transhumanism rather than being a philosophy or school of thought is better considered a social or political movement. Such activities, too, fall squarely within the ordinary spectrum of democratically legitimate forms of advocacy of most social, political or technological goal that are also employed by non-visionary, realist advocates of technology. Even more so, when we look closely at the "shaping" and "practice-orientation" of many technological utopians rather than being reckless toward technological progress, they utilize their thinking through the prism of a possible future to point out dangers and risk of technological change that they demand be mitigated. In Max More's outline of the "proactionary principle", we find that the principle suggests "boldly stepping ahead while being mindful of where we put our feet" (More, 2013b, p. 261). More underscores that "[e]xtinction risks, like other technological risks, point to the need for combining vigorous technological advance and wise decision-making". Bostrom in his *Letter* suggests: "I fear that the pursuit of Utopia will bring out the worst in you. Many a moth has been incinerated in pursuit of a brighter future. Seek the light! **But approach with care** [...]". And further: "I recommend you go easy on your paradise-engineering until you have the wisdom to do it right" (Bostrom, 2008, p. 5; own emphasis). In many other publications, Bostrom soberly outlines and assesses various existential risks that might be caused by the unmonitored and ungoverned emergence of advanced AI—superintelligence (Bostrom, 2016). There, he suggests a number of ways—albeit not very specific—of supervising and governing said technological developments, among others by supporting networks "comprising [of] individuals devoted to rational philanthropy, informed about existential risk, and discerning about the means of mitigation" and to carefully "manage information hazards" (ibid., p. 316).

Also relating to other authors, who speak favorably of technological utopias, such as Danaher and Chalmers, who overall cherish the benefits of virtual reality technologies, we can deem their engagement and advocacy as balanced: We learned about Danaher's warning of the possibility of locked-in states, when moving toward a utopia of automation and considered Chalmers warning that "the social and political issues that arise in ordinary reality will also arise in VR" (Chalmers 2022, p. 359). Taken together, I fail to see a target in these examples for the aforementioned charge of Segal, who refers to current technological utopians as the "men and women who care primarily about getting rich and playing off anxieties over the future" (2006, p. 69). Undoubtedly, there are technological utopians who are negligent of technological risks and whose "practical orientation" is rather destructive, reckless and financially motivated. Yet, most advocates of technological utopias—at least those, we have encountered—cannot readily be attributed with these features.

In passing, we have now distinguished between various forms of practical orientation among contemporary technological utopians, none appears to be at first sight directed at violence. Some are innocuous and remain in the realm of intellectual stimulation, furthering discussion, breaking the spell of our perception of the world as being without real alternative—those functions that many utopian scholars deem central to the utopian endeavor as we saw in Chapter 2. Others are a bit more practical and political active and go beyond mere intellectual stimulation: There are those that explicitly warn and caution of the possible dangers and risks ahead and suggest ways to mitigate them and those who carefully want to move ahead within democratically legitimate confines to invest into science and technology, do scientific and technological research and build communities and networks to increase (political) advocacy for their cause (McCray, 2013).

Nevertheless, even within the latter democratically legitimate ways of advocacy of contemporary technological utopias, there lurk real dangers that must be acknowledged: It is concerning indeed that in the process of envisioning and realizing a particular technological future and entering a particular technological trajectory—even within the confines of democratic legitimacy, very few influential people take the steering wheel. The concern that these utopias are elitist is difficult to shake and, therefore, problematic (Sand, 2019; Tutton, 2022). When looking at the forerunners of the Transhumanist movement, we find CEOs of big tech companies (Ray Kurzweil), university professors (Nick Bostrom) and other white, male, able-bodied individuals that come from affluent backgrounds and enjoyed a high education (Shew, 2022).[3] If—as ethicists, social scientists and technology assessors—we engage with those visions, we perpetuate their already exposed standing in the discourse. Rather than elevating some unusual, out-of-the-box narratives of our future—perhaps some non-technological narratives, in fact—those remain hidden, while we give credit to the already established narrative and thereby cement the existing discursive hegemony, if one wishes to call it that way. What holds true for Transhumanism, applies to many other technological utopias that currently attract much societal attention: space tourism and the space colonization, nanotechnology, quantum technology, etc. (Rubenstein, 2022). These visions have been rarely conjured by way of widespread societal discourse in which a diversity of viewpoints have been taken up and mingled into the picture. In epistemic terms, that might mean they have missed on tackling important values or envisioning pathways that would be more original and overall better. The values they represent might, therefore, not resonate with society writ large. And even if their implementation proceeds within the confines of democratically legitimate means, there is a risk of sidelining the concerns and values of less privileged populations, whose democratic power to influence technological change is indirect at best. I do not know whether there are neat and simple ways to get to the lesser known, more mundane futures and make them more visible in the debate, but I understand and acknowledge the risk of them being pushed outside the purview of academic attention and discussion (see also Groves, 2023; Sand, 2019).[4]

As mentioned before, I will do little to debunk this suggestion. Still, I would insist that we separate the utopias' content from their practice

[3] As I have suggested before, the picture is more complex when we delve into the varieties of the movement and their connections to, e.g. grassroots biohacking communities (see e.g. MacFarlane, 2020).

[4] Sargisson also points this out and suggests that it is important to investigate present utopias through the following questions: "These terms are authorship, context and intent. If we ask questions cast in these terms (such as, 'Whose utopia is this?' 'In what context does this occur?' and 'What does this seek?'), we can untangle some very different types of utopianism. Together those provide insights that are important for their evaluation" (2012, p. 240).

for analytical purposes. Furthermore, a primary objective of the present essay is and remains to underscore that we need to place these concerns next to several benefits that are often overlooked in this discussion and outweigh—from my point of view—those dangers. Additionally, we must not negate the at least partial responsibility that recipients of such visions have in not taking them uncritically for granted and immediately trying to realize them.

3.2 Engagement with Unrealistic Scenarios Deflects from More Important Present Concerns

This argument has been forcefully advanced by Alfred Nordmann and continues to entice intriguing scholarship (Rueda, 2023; Zohny, 2021). In *If and Then: A Critique of Speculative Nanoethics*, we find the following compelling view already in the article's abstract:

> This discourse [about speculative nanoethics and human enhancement] violates conditions of intelligibility, squanders the scarce and valuable resource of ethical concern, and misleads by casting remote possibilities or philosophical thought-experiments as foresight about likely technical developments. In effect, it deflects consideration from the transformative technologies of the present. (Nordmann, 2007, p. 31)

This is a rich criticism that needs some disentangling—it is a three-pronged attack against engagement with speculative technological futures. These technological futures undoubtedly also refer to the technological utopias that have been introduced before. My response will be to suggest that qua narratives we can choose to treat technological futures as predictions or as utopia. Only the former gives rise to the concerns that Nordmann outlines, but we can confidently omit such reading. From the outset, Nordmann's view is problematic, as I outlined elsewhere (Sand, 2016b), insofar as an assessment of technological futures as being either "remotely possibly" or in fact relevant for the "transformative technologies of the present" does require some form of study. Without at least a cursory glance at technological futures, there can be no determination about whether they are rather realistic, or merely remotely possible. So, the study of technological futures is important to even make the distinction that is invoked in Nordmann's argument. Still, the cursory glance will not amount to the engagement that we are looking to defend here.

Nordmann's first attack suggests that engaging with such technological futures means *casting the narrative as a prediction* of something that is a mere remote possibility. Perhaps, Nordmann has something like the following examples in mind, where utopians appear more as prophets; they predict the future rather than playing with ideas about it. Consider, these examples from Negroponte, who writes in *Being Digital* about the impact of computer communication: "The change from atoms to bits is irrevocable and unstoppable" (Negroponte, 1995, pp. 4, 229). This is clearly a prediction—and utopia does not square well with predictions. That would be rather paradoxical indeed: If a utopia is about thinking in alternatives and estranging so that a better world can be pursued in whichever way, then predictions are not the right place for utopias (Hausteiner, 2022, p. 129; Levitas, 2011; Saage, 2016, p. 64). These agents appear as prophets rather than as utopias—they are not interested in creatively envisioning different ways of living and being to entice relevant stakeholders—they suggest that there is little control over those developments anyhow: "Like a force of nature, the digital age cannot be denied or stopped", writes Negroponte (1995, p. 229). What value does it then have to paint a picture of the digital age? Of course, one could claim that just as the impossible utopias like the Land of Cockayne can incite hope or at least deflect from the hardship of the present, so could the idea of the irrevocability of the digital age as expressed by Negroponte incite hope in the future. Though, the lack of human control over those developments that is implied here, makes the ethical discussion that Nordmann is interested in, rather futile. Surely, there are many prophets among technological utopians, yet examples to the contrary suggest a more diverse landscape. In Max More, we have already found a Transhumanist, who is skeptic of the possibilities of prediction. James Hughes, too, underscores the limits of prediction: "Trying to predict the progress of these evolving and cross-fertilizing fields requires a good bit of humility" (Hughes, 2004, p. 7).

Of course, Nordmann is not concerned about the predictions themselves, but about our engagement with them: Do we become complicit in the act of forecasting, if we engage with those narratives as utopias? To me, this seems rather implausible. Consider the same act, but with reverse intent: Just as one could warn with reference to literary works such as *Brave New World* that some forms of technological surveillance have freedom undermining capacities and can be abused to hegemonize political power (Reiman, 1995), without claiming that any such technology is on the rise, one could engage and extract normatively relevant insights from technological futures without suggesting that those futures are within our grasp or realistically emerging in the upcoming years. The reflection on the likelihood of a certain technological future and an assessment of what social, technological and scientific conditions would have to obtain in order to realize such future, are part of a critical ethical assessment and stand detrimental to the idea of a forecast. The ethical engagement, in fact, suggests means of intervention and control over future technologies that many forecasts implicitly deny. What value would normative assessment have, if we could not at least to some degree shape the trajectories and designs of currently emerging technologies. Again, this position is advanced from the perspective of someone seeking normative guidance.

The second dimension of Nordmann's argument is an epistemic one that concerns the status of prediction and the value of the subsequent assessment—the normative consequences we can draw from an engagement with those visions (Zohny, 2021). He has in mind far-fetched ideas about Grey Goos (seemingly self-contained Nanoparticles that continuously reproduce to finally deplete all earthly resources, which is an idea made famous by Eric Drexler [1986]), mind-uploading and technological immortality. Oftentimes—in particular in the early days of nanotechnology—perhaps a more widespread approach to these scenarios was to consider what if this or that technology indeed makes us immortal; what would the ethical implications be, if this were to become a reality? Consider the following speculations by John Harris as an example: "We thus face the prospect of 'mortals' and 'immortals' existing alongside one another. Such parallel populations seem inherently undesirable, but it is not clear that we could, or should, do anything to prevent such a prospect for reasons of justice or morality. If increased life expectancy is a good, should we deny palpable goods to some people because we cannot provide them for everyone?" (Harris, 2000, p. 59).

Nordmann claims that such questions are highly flawed: It is both hubristic to believe that we could predict such fundamental societal changes and all its side-effects and that—even if we could—have *control over such future* (Nordmann, 2013). But do we have an alternative? Can there be a short-term or long-term assessment of the future so that one or a number of alleged elements are taken as fixed, while others change, and their change is meaningfully evaluated? Future predictions that home in only one aspect of society (such as the longevity of its members) appear as massive simplifications—they disregard how different societal institutions and individual behavior intertwine. A more plausible image would assume that various elements (social, political, technological) chaotically interact, which seems to negate the possibility for meaningfully assessing them. Nordmann argues that some narratives implicitly and without justification leave much if not everything of the world fixed. We are being asked, for instance, to imagine being immortal (see Harris' examples above). It seems perfectly clear that—once immortality becomes a reality—much more than this basic fact about human biology will have changed along the way, which makes questionable that concerns about justice, for example, will be raised in exactly the same way as they are raised now. Hence, in order to assess the morality of the state that has emerged, one artificially imposes a concept of justice from the present to future beings, whose concept is starkly different. Hence, current "if–then"-assessments of the immortality narrative implicitly assume that except for our mortal bodies much remains fixed in the future—including human desires, our concepts of human nature, our ideas of friendship and justice—and then jumps toward an assessment of that one aspect that we imagine having been changed; a flawed exercise, or so it seems.

But, first, is the demand for complete knowledge of a future state including its values and its many accompanied socio-political and moral drifts not excessive? *Prima facie*, it seems that one can establish assessments of various fragments of the present, while being ignorant of many other aspects of that same present. Although, it might clear that if someone were to plan an attack on a famous politician next year—that would be morally wrong. If it is so, then it is so under the assumption that that politician has not been planning to go to war with another country or commit genocide the following day. So, in every counterfactual evaluation of a situation (and perhaps any evaluation), implicit (factual and normative) assumptions are being made (Scheffler, 1993). Hence, this does not seem to be a feature of evaluations of futuristic visions only. Because of the necessary epistemic incompleteness, all evaluations have an ad hoc character. They commence assuming that in ordinary situations, a sufficient degree of information is available. Given the widespread nature of this ad hoc undertaking, and the plain lack of alternatives to this procedure (How would we know what future people's values and concept of justice are?), it is fair to say that it would be too much required to ask for a comprehensive picture of the world (how comprehensive anyhow?), if we would want to be able to assess some of its features, like human immortality within it. If we had to know everything about the world in order to evaluate certain fragments of it, there could be no evaluating. But given the need of prospective assessments, we must do with what we have at our disposal. A predictive theory of value change is unfortunately not among the available repertoire of tools (though big strides have been made toward a better understanding of the phenomenon of value change in recent years and its existence is now largely acknowledged [Danaher, 2021; van de Poel, 2021]). Rather than stopping to assess, we must be extra cautious and underscore the fallibility of the endeavor.

Furthermore, unlike perhaps fields like TA, where societal values are the yardstick for assessments of technology, in most of applied ethics, one does not actually invoke the values and normative convictions of present society—whatever they are thought to be: In their totality, they are conflicting and plural. Rather, one alleges and assumes a somewhat more objective normative stance. Why should we not be able to assume a similar normative stance for assessing the future. In this way, Nordmann's challenge to assessing the future already loses some of its threat.

Finally, again, the major problem, is that the idea is premised on the assumption that we must treat the narrative as a sort of prediction: There is no need to assume that a narrative provides orientation and guidance, if and only if, they can correctly predict the future. This seems to have been the cornerstone of Engels' critique of the early utopian socialist: Engels charges utopian socialists for being unscientific and resting their account on speculation and an incomprehensive understanding of the socio-economic system and its dynamics (Saage, 2019, p. 20). Nordmann would add that it cannot be any different than being unscientific, because there is no scientifically robust way to access the future anyhow—an idea that also Popper would readily agree with (Popper, 2012). However, utopian narratives including technological utopias need no scientific basis. They need *not* be understood as predictions about the future at all. They can provide guidance and orientation and creative input and initiate important discussions about the idea of justice, for example, without claiming to predict the future. Demanding else is a bit as if we were tasked to read a crime novel (e.g. *Crime and Punishment*) like a court briefing—a categorical mistake. We must read technological utopias less as roadmaps toward the future (albeit much evidence fuels such reading—as we have seen in previous sections), but more as abstract paintings that reveal new insights for every following generation studying them.[5]

Third, Nordmann suggests an intriguing resource argument: Utopias (or more broadly technological futures and visions in general) might deflect, as Nordmann says, from more urgent concerns in the present (Nordmann, 2007). How much force does this argument carry? In a strong reading, the idea of giving the present and present concerns overwhelming precedence over the future and future concerns appears irrational. Consider that many people forfeit at least certain goods to ensure the acquisition of future goods, which underlies insurance investments and saving decisions. In such common acts, we indeed give precedence to concerns that we expect to be lying in an uncertain future. We consider the likelihood of our continuous existence and anticipate the needs that we will most likely experience then. Then, we forfeit the consumption of goods in the present to ensure the enjoyment of other, perhaps greater goods in the future. Hence, a pure "presentism" of the kind Nordmann suggests that attributes overwhelming weight to present concerns and needs, does not attract much support or *prima facie* plausibility.

What exact weight a distant future possibility shall be attributed in present policymaking, is of course, not easily determined: How much should we invest to guard us against meteorite impacts? But, discarding *all* distant future possibility would be misguided, I suggest. Hence, the risk that we give too much weight to the future in a desperate move to try preempting it or—the other way around—further a certain future is prevalent, just as the risk that we give it too little weight to it and mistake a lack of complete control with lack of any form of control.

[5] Nordmann seems to agree with this when writing: "Indeed, if science fiction scenarios lead to interesting philosophical questions, it is precisely because one suspends disbelief in the presence of fiction. Relieved of the pressure to determine what is true or false, what is likely to happen and what not, we can forge ahead and explore who we are, who we might wish to be, and how these wishes reflect on ourselves or our views of human nature. In other words, there is nothing wrong with public debate of human enhancement technologies or molecular manufacturing where such visions provide a backdrop for society to reflect upon itself" (p. 43). My argument merely suggests that this can be done, while completely ignoring the predictive elements that are usually conjoined.

Furthermore, we must see this argument in context in which it has been uttered and to whom it has been uttered: It has been directed at the ethics community at the time when nanoethics flourished. That was perhaps a time, where—for the purpose of receiving third-party funding—many ethicists underscored the risks of emerging nanotechnologies without knowing much about them and, perhaps, without believing in them, thereby perpetuating these visions (Sand, 2016a). They might have become complicit in a game of underscoring the potential of an emerging technology to propel and emphasize the need for their own kind of research on those technologies. Perhaps the same is happening with AI now. However, how broadly applicable is the argument? As professional ethicists, political philosophers and utopian scholars, we all work in institutionalized settings, need to apply for grants and funding, which sometimes require us to underscore the societal need of a certain type of ethical investigation. This usually is done by outlining the possible consequences of certain technologies—sometimes exaggerating them for greater impact. In this sense, we are never completely apolitical, as Adam Briggle reminds us (Briggle, 2020). Still, these fields are characterized by a vast distribution of scientific labor: When we look at the case of Transhumanism, we see not only a variety of ways to engage deeply with the vision that are rather diverse; scrutinizing the defensibility of liberal eugenics (see Chapter 5), the history of the transhumanist idea, the eschatology of the view, their view of human nature, the relationship between post- and transhumanism. This division of labor in various academic fields makes implausible arguments that suggest that all of these forms of engagement contribute to the hegemony of a technological future and furnish a divergence from present concerns. Given the breadth of philosophy of technology (let alone Utopian Studies and political philosophy) as a professional discipline, aside from a diversity of engagement with more speculative visions, there can and should always coexist forms of research that concern the more short-term developments of technologies in the present. This can be done simultaneously—as my own work about the near-term consequences of medical AI and its ethical implications suggests (Sand & Jongsma, 2022). As in the realm of political theorizing, both engagement with realistic and idealistic proposal are simultaneously possible (Valentini, 2012).

3.3 Technological Utopias Fetishize the Technological and Are Socially Oblivious

The last charge that shall be discussed here, is equally intricate and comes in various flavors. We can read this first as a form of naiveté: The technological utopian who fetishes technologies as silver bullet solutions to all kinds of natural and societal problems overestimates the limited power that technology has to overcome socio-political challenges and problems of injustice. In a second reading, the argument suggests that technological utopianism is lacking the subversiveness—the criticality and brazenness that help us identify the deficiencies of the currently reigning social order so as to move forward toward a better way of living and being. In its acceptance of the liberal capitalist framework, technological utopias have lost their ability to provide a true alternative, and, therefore, lack one of utopias key characteristics and benefits—namely that they identify deficiencies in our current way of living by presenting narratives and which these have been resolved or are inexistent. *Prima facie*, it must count as a defect, if technological utopias lack what makes other utopias valuable and worthy of engagement. Even worse, if that were a necessary feature of utopianism, such technological narratives would not deserve carrying the title utopia.

Hence, technological utopias have oftentimes been criticized for fetishizing the technological and forgetting about the social, which contains severe risks. This can be understood in two ways, both of which shall be discussed in this and the following sections: (1) Technological utopias are socially naïve and, in this sense, *truly unrealistic*. Winner suggests regarding the vision of a comprehensively computerized society that "[w]hat one often finds emphasised though, however, is a vision of drastically altered social and political conditions, a future upheld as both desirable and, in all likelihood, inevitable. Politics, in other words, is not a secondary concern for many computer enthusiasts; it is a crucial, **albeit thoughtless**, part of their message" (Winner, 1986, p. 102; own emphasis). Winner goes on to clarify that "[if] the solution to problems of illiteracy and poor education were a question of information supply alone, then the best policy might be to increase the number of well-stocked libraries, making sure they were build in places where libraries do not presently exist" (p. 109). In short, he believes that there are better "political fixes"—one might exaggeratedly and pointedly call them—for the problem of limited access to and low quality of education. Kerrie

Jacobs notes more broadly and poignantly that the "world on this side of the computer screen is such a seamless continuation of the world on the other side that even the Secret Service is here" (Jacobs, 2001, p. 350). And Ted Peters suggests in his critique of Transhumanism that technology is always only as good as the humans that develop and utilize it. Given the corruptibility of humans, we cannot expect them to do any better with advanced technologies in their hands than without: "My thesis is this: transhumanist assumptions regarding progress are naïve because they fail to operate with an anthropology that is realistic regarding the human proclivity to turn good into evil" (Peters, 2011, p. 148). So, must politics save us from technology, or rather the other way around?

Surely, when we consider policy-decisions such as investments in science and technology, we must take feasibility into account. In previous works, colleagues and I have warned of falling for the trap of naiveté, when assessing technologies' potential: Real immortality is not a feat that material beings can achieve. This, as we have argued, has significant impact on some of the arguments on the "tedium of immortality" but also on the value of living very long lives; it implies that we cannot eradicate the grief of losing loved-ones, we can only postpone it (Sand & Jongsma, 2016). Postponement of grieving such loss might be a worthy goal, but such state shall not be conflated with immortality; that would be naïve. First, however, the contentious examples of technological utopias discussed here (Transhumanism in its many other ambitions, VR and computerized communication) advocate by and large goals that are less obviously impossible than that of metaphysical immortality. Second, it is important to underscore that the gravity of this objection stems from a reading of technological utopias as blueprints that ought to directly affect socio-technical change. The neglect on which critics put their finger here—the overvaluation of technology against other social or political forces—would be particularly troublesome if we were to utilize those narratives directly as a backdrop for present governance and technological decision-making. We would simply waste a lot of resources to try and "techno-fix" problems that either have socio-political origins (e.g. lack of incentives for politicians elected for four years to invest into the long-term future of humanity), or that thereby create problems of a bigger magnitude (furthering inequality and manmade climate change).

But, as I have argued before, we must not read technological utopias as blueprints or signposts in this way (even though they are often intended to fulfill this role). While it would certainly lead to massive disappointment and waste of resources on the way to embark on the path to immortality, pondering a world of immortal human beings might be interesting enough to understand better the meaning of life of mortal beings and how such meaning can be achieved. In painting such a world, the technological utopian initiates an important discussion and a process of reflection that might not be insignificant, albeit not of immediate practical relevance. To me, it is, therefore, remarkable that the charge of naiveté is leveled against technological utopianism by authors such as Thaler, Segal and Chrostowska,[6] who articulate pro-utopian sentiments throughout their work and should be sympathetic to the arguments that have been presented before. During his captivating discussion, Thaler suddenly applies a different standard of assessment, when he criticizes ecomodernists' faith in the redemptive power of science and technology, while no such standard is applied when he discusses other utopian outlooks such as the Gaia hypothesis. He writes:

> The problem that surfaces in this context is that, by highlighting optimistic themes in their visions of the future, ecomodernists, especially of the market-friendly variety, tend to think about human ingenuity in a rather simplistic manner, as the sole font of unblemished progress. The violent and vicious side of our species is usually brushed away as an atavistic remainder of an age before the 'humanitarian revolution.' This tendency to overemphasize our species' resourcefulness becomes rampant in the ecomodernists' treatment of cutting-edge technologies that are intrinsically ambivalent. (2022, pp. 172 f.)

Thaler thinks that the "what-if" plotline that undergirds that Gaia hypothesis focuses rather on estrangement as a cognitive device than on the question, what practical (climate) actions such speculations about Earth as a living being should entice. He also criticizes this given its possible fallout of resignation and inertness as a typical shortcoming of "what-if"-plotlines. The "if-only"-plotline attributed to ecomodernist speculations has a more galvanizing ambition, according to Thaler. Hence, he might

[6] In footnote 2 in Chapter 1, we have introduced Segal's idea that in some modern technological utopias, technology becomes an end in itself. Chrostowska charges the technological utopian—Transhumanists in particular—of escapism and suggests that they ignore "emancipation's ethical, political, and economic dimension" (2021, p. 21).

consider that given this ambition, it might be more suitable to scrutinize the feasibility of the "if-only", but not that of the "what-if" plotline. However, one cannot completely distinguish the cognitive and dispositional functions of utopianism as suggested before (Sect. 2.2). This makes it untannable and, in a sense, unfair to pit an evaluation of feasibility against technological utopias, but not social utopias or those—like the Gaia hypothesis that stands completely outside that dichotomy. There is clearly something that each of these visions want to achieve, rather than being a mere stimulant of new perspectives, and it seems unfair to wonder whether that practical goal—whatever it is—is within the reach of feasibility for ecomodernists but not within reach of those who joggle with the idea of attributing agential features to Earth. If we are willing to affirm that non-technological visions (such as the Gaia hypothesis, or degrowth) can have positive theoretical and practical power, while being far from realistic, why should we measure technological utopias according to a very different anti-utopian standard—namely realism? So far, our counterarguments might only sway those, who already have pro-utopian leanings. What about realists (like Winner and Morozov), who endorse the view that technological utopias are naïve for being socially oblivious, overestimating the fallible humans behind the technologies? What do we have to offer them?

Again, we must wonder whether the alternatives to technological utopianism are free from the alleged defect. There is a good reason why technologies are suggested as solutions to a wide range of environmental and societal problems: The alternative—the allegedly realist policies and social transformations that have been proposed and advocated oftentimes over various decades or centuries such as Winner's idea of "simply building more schools and libraries"—has proven just as ineffective to resolve persistent problems such as inequality stemming from access to education. The lack of motivation of relevant stakeholders to support such initiatives, poses a persistent obstacle to resolve this problem ubiquitously. Who, then, is the real utopian here—those who clinch to the possibility of betterment through human action and politics or the socially oblivious technological utopian? Even rather "realistic utopias" (Wright, 2012), including the Universal Basic Income that have been widely advocated over the past decades to alleviate inequalities and remove the pressure and denigratory procedures from those in search for work, have not found widespread alliance. Various promising policies to halt climate change or mitigating its effects, such as Climate Clubs to establish monetary incentives for climate action, have not been adopted either (Nordhaus, 2015).

Richard Saage suggests that "any authentic utopia must under the Damocles sword of humanities potential to technologically self-destruct (through nuclear weapons) cannot further appeal to technological solutionism, but must convince humans to become materially more parsimonious" (Saage, 2019, p. 23). However, it is this exact same human motivation that in most cases through severe inertness toward reasoned change and sometimes by outright misconduct and moral corruptibility that brought us toward the brink of potential extinction through climate change and nuclear warfare and that causes the persistence of capitalism in the beginning of the twenty-first century. Hence, why should we and, in turn, utopians continue placing trust in our power to sway humans and redirect their inclinations by appeal to reason? On a general level, this seems just as unreasonable as maintaining one's hope—the utopian desire—in the redemptive powers of technology. What general conclusions can be derived from this?

To diametrically oppose the social, the political and the technological as many of the aforementioned authors do and argue that utopias must tackle both or rather focus on the former, is in my view to misunderstand their relationship. Not only could technology provide solutions to societal problems, but their introduction also creates new socio-political constellations that make upholding a radical distinction between the social and the technological realm untenable. To render some socio-political solutions redundant for societal problems by way of fixing them technologically, is central to the desire that technological utopians sustain. Less radical, however, is their impetus to re-envision and underscore the permeability of this relationship.

A realistic view suggests that currently, technology and politics are intertwined and neither has the prerogative as the primary target to initiate socio-political change. To many, the former half of the sentence seems very much agreeable (Hobson & Roessing, 2022), while the latter appears to be more controversial and finds less support: Authors, like Peters and Morozov believe that technology is only as good as the politics that ultimately domesticate it. Morozov suggests that politics is in fact more important than technology and considers the opposite position explicitly as a form of naiveté—an unreasonable optimism in the powers of technology to shape politics rather than the other way round:

> Instead [of cyber-utopianism], we'll need to opt for policies informed by a realistic assessment of the risks and dangers posed by the Internet, matched by a highly scrupulous and unbiased assessment of its promises, and a theory of action that is highly sensitive to the local context, that is cognizant of the complex connections between the Internet and the rest of foreign policymaking, and that originates not in what technology allows but in what a certain geopolitical environment requires. (Morozov, 2011, p. xvii)

We have suggested before that it is odd to argue against utopianism as a form of naiveté. That is in essence saying not much more than that they are utopian, and it is not obviously clear that that is a defect (as long as we do not additionally claim that they need to be realized [immediately]). We should not consider either technology or politics as the primary instrument to change society. Very few commentators seem to be willing to consider that we must perhaps focus on politics and technologies alike. Like Morozov, they look at the dangers of trusting technologies blindly. As evidence, they cite an ever-increasing list of adverse environmental and societal impacts that technologies had on the planet, on non-human and human animals. Due to these failures (intentional in the hands of dictators and ill-meaning humans, and unintentional ones such as the environmental crises in its early inception), they prefer to focus on politics. But, here too, we could turn the table and instead of focusing on the failures of technology, we could look at the failures of men: slavery, colonialism, the Holocaust, the lack of concerted efforts to mitigate climate change. Such horrific atrocities have been done *with* technologies, not *by* them. Hence, we could draw the exact opposite conclusion and utilize the evidence of mankind's crookedness to refuel our hopes in technology instead of politics. *Prima facie*, neither of the two can claim the upper hand here. When Sherry Turkle complains in the subtitle of her *Alone Together* that in modernity "we seem to expect ever 'more from technology and less from each other'" (Turkle, 2011)—it appears all too easy to point the finger at the causes of this faltering trust; the crimes against humanity and other atrocities of colonialization and climate change that have been mentioned before are but a few.[7]

[7] "Viewed from Anders's perspective, therefore, the enlightenment no longer looks like a project of human emancipation and democratisation that leads to the rise of social institutions that seek to address human inequality. On the contrary, it is a project which, on a more fundamental level, represents an active turn away from, and a faltering trust in,

Politics and technology both shape and impact another and are shaped and impacted by another (Nye, 2006). There are roughly four ways of how politics and technology can interact: (1) Technologies can *enhance the impact of political decisions* and regulations, which is desirable, if these decisions and regulations themselves are desirable. Speed cameras, for instance, monitor speeding behavior. Their presence helps to discourage and prosecute traffic offenders. Naturally, technologies can also enhance the impact of bad policies and regulations. But, (2) technologies can also *undermine the impact of political decisions* and regulations, which is desirable, if those are bad decisions. One can here think of leaflets or other communication technologies that are used to spread information that autocratic regimes aim to suppress. (3) Policies, on the other hand, can curtail technological developments or *undermine their effects*, which is desirable, again, if those technological developments are bad. In this way, European countries have, for instance, started to regulate vaping. Lastly, (4) policies can also *stimulate and enhance* such developments, if these are deemed good. Investments and subsidies into sustainable technologies in the past years are prime example. What is seen as desirable or undesirable is, of course, often a matter of contention. Figure 3.1 shall display this reciprocal relationship.

When talking about the basic structure of society in more detail in the following chapter, we will get a clearer picture of the interrelation between technology and socio-political institutions in relation to theories of justice. We will insert individual behavior into our picture of location of justice to form a triangle. We will see that utopias re-envision how technologies, individual behavior and socio-political institutions could possibly interact in the future, and how, thereby, new socio-technical constellations emerge in which justice is realized or at least more approximated. In this triangle toward a just society, as will be established, we will see that technological utopias strongly accentuate the pillar of technology and sometimes envision the submergence of the other pillars under the direction of technology. Pondering the latter possibility falls squarely within a tradition of utopianism that have accentuated either individual

everything human'" (Anders & Müller, 2016). This faltering trust that Ander's detects, as I argue above, shall not only be seen as a byproduct of the existence of technology as an alternative means, but also of the persistence of evil and corruptibility of human individuals.

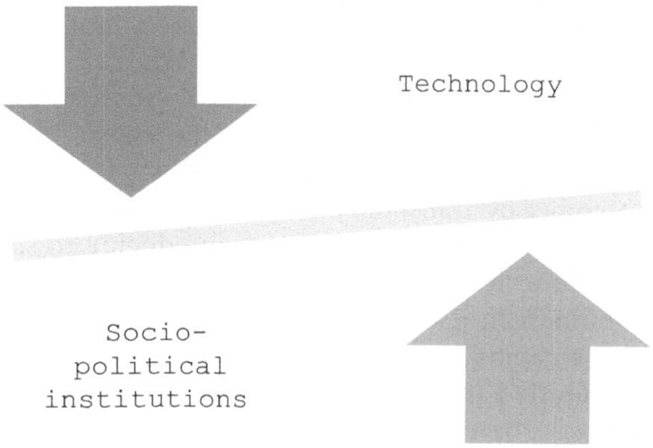

Fig. 3.1 Reciprocal relation between technology and socio-political institutions

behavior (ethos) or socio-technical institutions as *primary* loci for their envisioned transformations.

3.4 The Social Obliviousness of Technological Utopias Amounts to an Acceptance of Liberal Capitalism

There is another, a second way to unpack the argument that technological utopias are socially oblivious: The suggestion is that the individualism and seeming endorsement of liberal markets that underlies many contemporary technological utopias and, in particular, Transhumanism, omits presenting an actual alternative to our present society. These technological utopias, therefore, loses much of the appeal that characterizes classic utopias that provide real alternatives to the reigning socio-political order and, thereby, unfold their subversive power. The general sentiment is well expressed by Dickel and Schrape, when they write about contemporary digital utopias that "[t]hey focus on the transformation of the material environment of human existence or human nature itself. Contrary to traditional social utopias, which wed social criticism and alternative conceptions of society, in the scheme of technological utopianism **society**

exists merely as background noise" (Dickel & Schrape, 2017, p. 47; own emphasis).[8]

For various reasons, Richard Saage warns of the Transhumanist takeover of utopianism. He suggests that "if the classic, political utopia intends to have a future, it must find ways and means to successfully resist the takeover attempts by transhumanist improvements in man" (Saage, 2013, p. 6). And further, he charges "those who decide to subsume both approaches under one term 'utopia', blur its boundaries and abandon it to such arbitrariness in its application in concrete phenomena that is comparable with a capitulation of the transhumanist challenge" (ibid.). As I am not committed to a particular notion of utopianism and think that those boundaries have always been more blurred than here assumed and continue to blur further, I am obviously not quite as concerned as Saage about this conceptual expansion. In the previous chapter, I followed willingly authors who include Transhumanism into utopias' canon. Interestingly, Saage assumes here that utopianism needs defense. Utopianism has an essential core, whose purity needs to be conserved and protected. This is not defended in the article, and we have earlier exposed the myriad, diffuse and diluted meanings that are now associated with the concept—and perhaps always have been associated. If instead, we look at the question "What makes utopia valuable?", we might find features that Transhumanism shares with other utopias, even if it does not share all aspects of "traditional" utopianism.

What makes Transhumanism so threatening in Saage's view? The alleged minor role that societal critique plays in that utopian narrative is underscored in various publications: "The 'society' plays only a marginal role in providing the material for the framework of the innovative engineer, who expedites technological progress in a certain social situation against possible constraints" (Saage, 2013, p. 4). He continues later to suggest that "[Transhumanists] do not change the socio-political set-up in which they originated. On the contrary: the 'new' society is interpreted as

[8] It is important to note that I am here presenting an abbreviated passage from Dickel and Schrape (2017), who immediately after express a view very much in line with my own: "[T]he utopian discourse of today can no longer be reduced to a polarity of technologically oblivious *social utopias* versus socially oblivious *technological utopias*. In the internet age, exceedingly popular *media utopias* combine—at first sight free of any ideology—expectations of technological potential and far-reaching ideas of social transformation overlaid with a shimmering revolutionary vocabulary in a novel and particular way" (p. 47).

the product of transhumanist innovation, for e.g., in the categories of the 'information society' on the unchanged basis of its capitalistic and national society of origin, whose market-conforming interest is what it explicitly aims at" (Saage, 2013, p. 4). Altogether—assuming as we have established before that technological utopias cannot and should not be practically relevant in a narrow sense—this would mean that they cannot even fulfill the cognitive function, hence be practically relevant in a broader, derivative sense. In summary, they might *not* be worthy of engagement. The argument might be sketched as follows:

1. A utopia is valuable and worthy of engagement if and only if it sketches a real alternative to reigning socio-political order and, thereby, critically scrutinize our current socio-political order.
2. Technological utopias á la Transhumanism take the reigning socio-political order as (liberal capitalism) as a given and accept the individual consumerist desires that have resulted from it.
3. Taking liberal capitalism with its resulting consumerism and individualization tendencies as an unchangeable given, nullifies the attempt of offering a real alternative.
4. Therefore, either technological utopias á la Transhumanism are not worthy of being called "utopias" or they are not valuable and worthy of engagement.

Both conclusions vitiate the overarching goal of this essay—at least in relation to Transhumanism as an example of a technological utopia. However, other technological utopias might be prone to such criticism, too. Hence, the argument is a challenge and deserves careful discussion. The following discussion paves the way for our examination of the scope and location of justice in the next chapter. In the argument, Saage virtually equates the notion of criticism (subversiveness), whose role for utopianism we readily acknowledged, with criticism of the reigning socio-political order. He identifies liberal capitalism as the reigning socio-political order. He clearly narrows what utopia ought to envision as a better way of living and being.

At first sight, Transhumanism seems like a fairly good target for such criticism: In his *Letter from Utopia*, which has already been introduced, Bostrom asserts cryptically: "To be sure, Utopia is not a location or a form of social organization" (Bostrom, 2008, p. 6). He seems to negate the need for any form of social organization in a posthuman world. His strongly individualist outlook is pronouncedly expressed in another famous article, *Why I want to be posthuman*, where he outlines his key claim as the idea that "for most current human beings, there are possible posthuman modes of being such that it could be good for these humans to become posthuman in one of those ways" (Bostrom, 2013, p. 42). That a posthuman, enhanced mode of being would be good for many *individuals* leaves remarkably open, how such a society would look like and whether it would be good *for that society*. The deficiencies that Transhumanists usually detect are related to individual abilities and their natural makeup: How fast they can run, how well they can hear, how long they can live and what they can experience. Whether or not, for instance, particular social groups are able to participate in political decision-making, is nowhere expressed as a prominent concern. It is, therefore, also unsurprising that most of the objections that have been leveled against Transhumanism concern how a society with only posthuman beings, or a society where "ordinary" humans and posthumans live side by side, would look like and be organized. Naturally, one wonders whether massive inequalities would ensue, if some beings were radically enhanced and others not (Kass, 2003)? The interaction and social organization of those posthuman individuals is hardly ever explicitly addressed, also not in his *Letter from Utopia*. It is primarily outlined and underscored what great gains these beings *as individuals* would reap, if they were to become posthuman. So, the individual stands in the fore of the philosophical outlook of Transhumanism—society almost invisibly in background.[9] Additionally, as one might concede, many technological utopias are blind to the systemic, societal origins of the individual deficiencies they describe and want to overcome. While Transhumanist James Hughes acknowledges the societal circumstances and inequalities

[9] James Hughes, a defender of "democratic Transhumanism", would probably strongly oppose this reading. He also suggests that believing in the power of technology is not the same as abiding to capitalism: "Left Luddites inappropriately equate technologies with the power relations around those technologies, and try to fight capitalism or patriarchy or hierarchy by fighting technologies instead of by liberating the technologies for free and equal use. Technologies may make certain kinds of power more likely than others, but they do not *determine* power relations. Each new technology creates a new terrain for organizing and democratic struggle, new possibilities for expanded liberty and equality, or for oppression and exploitation. Technological innovation needs to be democratically regulated and guided, not fought or forbidden" (2004, p. 195).

that cause vastly different life expectations (Hughes, 2004, p. 24), Aubrey de Grey is primarily interested in a biological reduction of aging and longevity (Sand & Jongsma, 2016).

Altogether, this seems to support Saage's point that Transhumanism—and perhaps other contemporary, technological utopias—are in fact devoid of the criticality that is so central to the classics of the genre and fail, therefore, to fulfill the important function that traditional utopias fulfill; being subversive and estranging from the present so that its ideological spell can be broken. But, does this warrant the conclusion that Transhumanism—as a utopia—is entirely devoid of value and, therefore, not worthy of engagement? Such reasoning is problematic in at least two ways.

On the one hand, Saage is neglecting vast differences within liberal capitalist societies, when insinuating that the label refers to *one* very specific form of socio-political and economic organization.[10] This ignores the significant differences among societies that could be attributed with this label: the US, Japan, the UK, Poland, Brazil, Germany are all liberal capitalist states of very different guises. Each of these countries have radically different ways of organizing markets, providing health care, social security, fostering traditions, of regulating work and traffic and protecting their citizens, leading to vastly different lived realities and experiences in those countries. This suggests that even within the liberal capitalist framework radically different ways of living and being better are conceivable. In fact, a plausible case can be made that considers those liberal capitalist countries as still falling short of attaining their own political ideals: Recent austerity politics in the UK and the US constrain the free exchange of goods in global markets. Many liberal capitalist countries are further than ever from privatizing health care (e.g. Germany and the UK). The possibilities of exercising central liberal political rights such as that of voting are far from being available to everyone in the US (Dunn, 2014). Altogether, this suggests that liberalism itself could be seen as a utopia, whose lived realities are as far off from the ideal of liberal theory as the Soviet Union was from the ideal of communism (Gray, 2008). As Charles Mill says: "Liberalism as ideal turns out to be illiberalism in actuality" (Mills, 2020, p. 4). This, in short, suggests that there can be a *liberal capitalist utopia*: There is room for (more or less radical) change toward a better

[10] Which, of course, begs the question: What exactly makes a society liberal capitalist?

way of living and being even *within* the socio-political system currently attributed with that concept. This appertains not only to radical technological transformations: A feminist utopia of gender equality or a utopia of environmentally friendly economic growth (ecomodernism) can, with good reasons, be seen as idealistic and far off reality without immediately shaking the roots of liberal capitalism. Why not call a liberal-ecological vision—like the aforementioned idea of Climate Clubs that ought to be established *within* the current global market order—utopian (Nordhaus, 2015)? Erik Olin Wright has provided various radical ideas that would fundamentally change how labor and property are seen without challenging the general liberal capitalist outlook of our current societies (Wright, 2012). Transhumanism might neatly rank among those.

Saage thinks that not only is there no structural-institutional criticism being made in the current technological utopia but implies that there is no social criticism being made here at all. He, thereby, defines the concept of social criticism much narrower than I previously did, when I suggested that by presenting us with a better way of living and being, utopias estrange us from the present. The presentation of an alternative to our present way of living and being reveals flaws in the present makeup, galvanizes change and provides creative input about where to go: Together, this constitutes utopias' critical potential. Saage, in contrast, implicitly suggests that social criticism must today be a criticism of liberal capitalism. But that is anything but obvious. We have seen that all political theories have a utopian element to them: All of them can in some sense claim that they are and have never been properly realized. Lyman Sargent writes approvingly that all political ideologies have produced their own utopias—including liberal capitalism (Gray, 1994; Sargent, 2010).

Few liberals appreciate this thought: Joachim Fest complains that liberalism tolerates human fallibility and imperfection and differs in this respect fundamentally from the necessarily totalitarian outlook of utopia, which—as he suggests—entails the wish to produce fundamentally "better people" through suitable social institutions (Fest, 1991). Do liberal societies really live by that outlook—can they? It seems rather obvious that liberal societies accept the fallibility of humans only to a limited extent. To ensure a modicum of socio-political stability, many contemporary liberal states presume rather extensive ideological and ethnic homogeneity, which is contingent or enforced. Where this homogeneity was and is missing, the fragility of the liberal ideal emerges. The utopian character of liberalism—and its aspiration toward equality before the law—becomes particularly clear in the global context, when liberal democracies are confronted with alternative political systems (e.g. China, Islamic states) and cannot resolve these conflicts with their own democratic means. Democracies' first violent appearance in the French Revolution might count as an example of this general idea, as do more recent attempts to forcibly "export" American liberalism to the Middle East (Žižek, 2008). In these examples, human fallibility is by no means accepted, which Fest believes to be a central conviction of liberalism contra utopian political theories (Fest, 1991). The requirement of formal equality before the law and its implications for the organization of political representation is then imposed, and not with democratic means, because using these would most likely lead to a different form of political organization. John Dunn has extensively outlined that American style liberal democracy is an ideal that never was a reality (Dunn, 2014). This, on the one hand, allows us to interpret the (unfulfilled) global promises of prosperity of capitalism and liberal democracies as utopian. On the other hand, nothing fundamental would stand in the way of interpreting, for instance, the technological utopia of Transhumanism as a socially critical narrative—even *if* we were centered around liberal capitalist convictions. Even if we were to understand, thus, the—occasionally extreme—neglect of questions about just distribution and societal problems in Transhumanist writings as evidence for its liberal capitalist spirit, we must not deny them the label utopian or downplay their potential for subversiveness and estrangement.[11]

Another understanding of this general criticism homes in on the excessive individualism of Transhumanism which supposedly undermines its subversiveness.[12] Saage suggests that the "classic utopia was from the

[11] How widely the liberal capitalist spirit has been inhaled by Transhumanism is, of course, a contested matter: My argument does not require an answer to this. It can commence on a strong reading. But, here as elsewhere, critics of Transhumanism run into the risk of shooting at a strawman, given the many facets of the movement. Hughes (2004) acknowledges the variety of political flavors in Transhumanism: "Some transhumanists are libertarians, closer to the Wired, cyber-libertarian, technophile subculture, dismissing risks and arguing for a free market in genetic enhancements and nanotech. Other transhumanists, like Buchanan and colleagues, favor a social welfare state, with public health regulations, national health insurance and efforts to maintain a rough equality among humans even as they upgrade themselves" (p. xv). I have organized my arguments in relation to the works of some of Transhumanism's most serious proponents: Nick Bostrom, James Hughes and Max More.

[12] Saage has expressed this already a decade before in a German publication: "In den Neuen Utopien dagegen wird den Individualisierungstendenzen der Herkunftsgesellschaft

beginning **anti-individualistic**: the whole always had priority over the part or individual" (2016, p. 59; own emphasis). He might put the point like this: Individualism cannot form the center of a utopia that truly strives for a just and ideal society. At least, it would not fall within the canon of traditional utopias. The traditional utopias think about societies as a whole—they are collectivist or communitarian in spirit.[13] As Hauskeller notes: "[…] the transhumanist account of posthuman existence is an obvious wish-fulfillment fantasy" (Hauskeller, 2012, p. 41). One might read this as another flaw of the Transhumanist utopia, the absence of societal criticism. It seems that Transhumanists fancy a society very much like our own—but exaggerated in its acclamation of individual (viz. anti-communitarian) flourishing and wish-fulfillment.

When it comes to the idea of justice and its location, as invoked here, the previous arguments *foreclose*, where to locate justice. They suggest that the locus of justice lies primarily on the level of collective decision-making and socio-political institutions. We need more strictly to separate who is the addressee, *the agent of change*, and *for whom* do we instantiate such change for reasons of justice. These are crucially conflated in Saage's critique. The Transhumanist envisions perhaps first and foremost a change toward a posthuman state of being that is desirable for the individual. The agent of change seems to be the individual (although it might be better—as we shall later see—to identify technology as the key locus of the transformation). On the one hand, however, that does not rule out that said change is also associated or strongly accompanied by means of socio-political transformations—such as making available certain technologies for a large populations by way of distribution and cost reduction, etc. (Hughes, 2004, p. 9). On the other hand, the technologically enhanced

voll Rechnung getragen. Die Selbsterfindung des Neuen Menschen findet im Medium seiner Subjektivität und seines Wohlbefindens statt". He contrasts this: "Der Neue Mensch [in der traditionellen Utopie] ist vor **allem ein kollektives Wesen**, eindeutig bezogen auf den idealen Staat, in dem und für den er lebt" (2007, pp. 613 f.; own emphasis).

[13] "For about fifty years, from roughly 1920 through 1970, technology was commonly conceived as being the solution to large-scale problems such as poverty and education. In recent decades, by contrast, technology has increasingly come to be associated with the fulfillment of individual needs and desires, ranging from online college education to virtual travel to cyberspace relationships. High-tech provides choices in an ever more digitized, programmable world. **It facilitates a consumer mentality that asks what a specific device will do for oneself rather than for society as a whole** […]" (Segal, 2006, p. 68; own emphasis).

individual, even if we were to see it as the primary agent of change—must not at the same time be deemed the sole object of change—the entity for whom change is instigated. Furthermore, neither does the individualist focus logically rule out that said change of the individual is at least equally if not more desirable for society as a whole. Transhumanist James Hughes pointedly brings to the fore how an increase of individual well-being and the transcendence of individual capabilities, which forms a central tenet of the Transhumanist philosophy, could affect social and communal living and then reversely affect individual well-being: "Freeing people from pain and depression and making them happy, cheerful and optimistic, may also encourage engagement with life, community and democracy. Democracy, in turn, is one of the best guarantors of general happiness of a population" (Hughes, 2004, p. 50).[14]

To take another example: If one believes that unconscious biases and cognitive deficiencies contribute to the emergence of some forms of social injustice, then correcting these individual deficiencies would resolve at least one cause of social injustice. These are central assumptions in the "moral enhancement" debate (Persson & Savulescu, 2012). In this debate, it is alleged that limited altruism toward distant strangers and the prevalence of *status quo* biases that undervalue more distant futures, is ingrained in human nature, who used to live in smaller communities before, where flourishing was possible based on communal solidarity. Such features, it is assumed, are fatal in a world in which technologies have long-term impacts on the environment and future generations. In his highly recommendable book *Inventions of Nemesis*, Douglas Mao counters Saage's assumption about traditional utopias' alleged collectivist spirit. Surveying numerous examples, he suggests that the distinction between individualist and collectivist utopias (or—borrowing from Fuz "utopias of men" vs. "utopias of measures" [p. 86]) might be a productive working hypothesis to organize historical material, but that many utopias tackle both dimensions with different accentuations. It seems, therefore, better to think of this divergence as a continuum rather than an exclusive dichotomy:

[14] For a critical view consider Agar, who suggests in his "second paradox" of technological progress that the rise of well-being of some individuals (through technology) is the root cause for the diminishing well-being of others and society as a whole (Agar, 2015). Neither of these passages prove conclusively what the right view on this relationship is—this stalemate opens an avenue for utopian contestation.

> If we divest ourselves for a moment of the urge to classify each utopia as either wholly managerial [...] or wholly transformative [...], we may quickly see that most utopias mix the two styles. [...] Yet different utopias mix these styles in different ways and in differing proportions. The managerial predominates in some; the transformative, in others. Thus we may do well to think of a utopian continuum, to consider that any given utopia roughly inhabits a point on a spectrum ranging from heavy reliance on techniques for dealing with people as it finds them, so to speak, to powerful investments in producing people of a new kind. (Mao, 2020, pp. 88 f.)[15]

Whatever the correct interpretation of the core of *traditional* utopias, we can summarize the main insights of the previous discussion as follows: We have not definitively proven that it is possible to arrive at an ideal or just society simply by addressing and transforming (in whichever way) its members. Though, we have seen that some Transhumanists fancy that idea. However, we must consider the thought that such is impossible as question begging. Deeming defective those technological utopias in which society plays only a minor role, means *predetermining* the question by whom and how societal problems or questions of social justice are to be addressed. This concerns justice's location. It fulfills a rather natural utopian function to incite the hope that at one point in time—due to technology or widespread human benevolence—we dispense of the need for socio-political institutions to govern the distribution of goods or individual behavior to ensure justice. And it would mean succumbing to the very ideological imprisonments that utopians aim to overcome, to keep on insisting that such change via a (technological) transformation of individuals' and their inclinations is impossible. Technological utopians persistently try to offset such unjustified realism.

References

Agar, N. (2015). *The Sceptical Optimist: Why Technology Isn't the Answer to Everything*. Oxford University Press.

[15] Consider this comment from Levitas: "In so far as utopias are alternative holistic models of society, they encourage us to think about the interrelationships of social processes. This helps to remind us that sometimes the most important social policies may be those that do not immediately look like 'social policy', but which bear on the whole structure of the (global) political economy and hence the social fabric" (Levitas, 2001, p. 450).

Anders, G., & Müller, C. J. (2016). *Prometheanism: Technology, Digital Culture and Human Obsolescence*. Rowman & Littlefield.

Beckert, J. (2016). *Fictional Expectations and Capitalist Dynamics: Imagined Futures*. Harvard University Press.

Bostrom, N. (2008). Letter from Utopia. *Studies in Ethics, Law, and Technology, 2*(1), 1–7. http://www.nickbostrom.com/utopia.html

Bostrom, N. (2013). Why I Want to Be Posthuman When I Grow Up. In M. More & N. Vita-More (Eds.), *The Transhumanist Reader* (pp. 28–53). Wiley-Blackwell.

Bostrom, N. (2016). *Superintelligence: Paths, Dangers, Strategies*. Oxford University Press.

Briggle, A. (2020). Philosophy of Technology as Politics. In S. Vallor (Ed.), *The Oxford Handbook of Philosophy of Technology* (pp. 191–210). Oxford University Press. https://doi.org/10.1093/oxfordhb/9780190851187.013.12

Chalmers, D. J. (2022). Reality+ : virtual worlds and the problems of philosophy (First edition ed.). W. W. Norton & Company.

Chrostowska, S. D. (2021). *Utopia in the Age of Survival—Between Myth and Politics*. Stanford University Press.

Danaher, J. (2021). Axiological Futurism: The Systematic Study of the Future of Values. *Futures, 132*, 102780. https://doi.org/10.1016/j.futures.2021.102780

Dickel, S., & Schrape, J.-F. (2017). The Logic of Digital Utopianism. *NanoEthics, 11*(1), 47–58. https://doi.org/10.1007/s11569-017-0285-6

Drexler, E. K. (1986). *Engines of Creation: The Coming Era of Nanotechnology*. Anchor Books.

Dunn, J. (2014). *Breaking Democracy's Spell*. Yale University Press. http://www.jstor.org/stable/j.ctt1bhkp0j

Fest, J. (1991). *Der zerstörte Traum - vom Ende des utopischen Zeitalters*. Siedler.

Gray, J. (1994). After the New Liberalism. *Social Research, 61*(3), 719–735.

Gray, J. (2008). *Black Mass—Apocalyptic Religion and the Death of Utopia*. Penguin Books.

Groves, C. (2023). On Profane Futures and Profane Futures Literacy. In A. Grunwald, A. Nordmann, & M. Sand (Eds.), *Hermeneutics, History, and Technology—The Call of the Future* (pp. 122–130). Routledge.

Harris, J. (2000). Intimations of Immortality. *Science, 288*(5463), 59–59. https://doi.org/10.1126/science.288.5463.59

Hauskeller, M. (2012). Reinventing Cockaigne: utopian themes in transhumanist thought. Hastings Cent Rep, 42(2), 39-47. https://doi.org/10.1002/hast.18

Hauskeller, M. (2014). Utopia. In S. L. Sorgner & R. Ranisch (Eds.), *Post- and Transhumanism: An Introduction* (pp. 101–108). Peter Lang.

Hausteiner, E. (2022). Radikale Netzwerke. Über die Notwendigkeit digitaler Utopien. In B. Klein & R. Schmidt (Eds.), *Was macht die Digitalisierung mit der Politik?* (pp. 123–136). https://doi.org/10.1515/9783110785265-010

Hobson, T., & Roessing, A. (2022). Questioning the Politics of Human Enhancement Technologies. In D. Sands (Ed.), *Bioethics and the Posthumanities* (pp. 53–66). Routledge.

Hughes, J. (2004). *Citizen Cyborg: Why Democratic Societies Must Respond to the Redesigned Human of the Future*. Westview Press.

Jacobs, K. (2001). Utopia Redux. In P. Ludlow (Ed.), *Crypto Anarchy, Cyberstates, and Pirate Utopias* (pp. 372–375). MIT Press.

Jameson, F. (2005). *Archeologies of the Future: The Desire Called Utopia and Other Science Fictions*. Verso.

Kass, L. (2003). Ageless Bodies, Happy Souls: Biotechnology and the Pursuit of Perfection *The New Atlantis* (1), 9–28. https://www.thenewatlantis.com/publications/ageless-bodies-happy-souls

Kumar, K. (1991). *Utopianism*. Open University Press.

Levitas, R. (2001). Against Work: A Utopian Incursion into Social Policy. *Critical Social Policy, 21*(4), 449–465.

Levitas, R. (2011). *The Concept of Utopia*. Peter Lang.

Mao, D. (2020). *Inventions of Nemesis: Utopia, Indignation, and Justice*. Princeton University Press.

MacFarlane, J. M. (2020). Transhumanism as a New Social Movement: The Techno-Centred Imagination. Palgrave (Springer).

McCray, P. (2013). *The Visioneers: How a Group of Elite Scientists Pursued Space Colonies, Nanotechnologies, and a Limitless Future*. Princeton University Press.

Mills, C. W. (2020). Theorizing Racial Justice. In C. U. o. N. York (Ed.), *Tanner Lecture on Human Values*.

More, M. (2013a). The Philosophy of Transhumanism. In M. More & N. Vita-More (Eds.), *The Transhumanist Reader* (pp. 3–17). Wiley-Blackwell.

More, M. (2013b). The Proactionary Principle: Optimizing Technological Outcome. In M. More & N. Vita-More (Eds.), *The Transhumanist Reader* (pp. 258–267). Wiley-Blackwell.

Morozov, E. (2011). *The Net Delusion: The Dark Side of Internet Freedom* (1st ed.). PublicAffairs.

Mosco, V. (2005). *The Digital Sublime: Myth, Power, and Cyberspace*. MIT Press.

Negroponte, N. (1995). *Being Digital*. Knopf.

Nordhaus, W. D. (2015). Climate Clubs: Overcoming Free-Riding in International Climate Policy. *American Economic Review, 105*(4), 1339–1370. https://doi.org/10.1257/aer.15000001

Nordmann, A. (2007). If and Then: A Critique of Speculative NanoEthics. *NanoEthics, 1*(1), 31–46. https://doi.org/10.1007/s11569-007-0007-6

Nordmann, A. (2013). (Im)Plausibility^2. *International Journal of Foresight and Innovation Policy, 9*(2–4), 125–132. https://doi.org/10.1504/ijfip.2013.058612
Nye, D. E. (2006). *Technology Matters: Questions to Live With.* MIT Press.
Persson, I., & Savulescu, J. (2012). *Unfit for the Future: The Need for Moral Enhancement* (1st ed.). Oxford University Press.
Peters, T. (2011). Transhumanism and the Posthuman Future: Will Technological Progress Get Us There? In W. Grassie et al. (Eds.), *H+/-: Transhumanism and Its Critics* (pp. 147–175). Metanexus.
Popper, K. R. (2012). *The Poverty of Historicism* (2nd ed.). Taylor and Francis.
Reiman, J. (1995). Driving to the panopticon: A philosophical exploration of the risks to privacy posed by the information technology of the future. *Santa Clara High Technology Law Journal, 11*(1), 27–44. http://digitalcommons.law.scu.edu/chtlj/vol11/iss1/5
Rubenstein, M.-J. (2022). *Astrotopia: The Dangerous Religion of the Corporate Space Race.* The University of Chicago Press.
Rueda, J. (2023). The Ethics of Doing Human Enhancement Ethics. *Futures, 153*, 103236. https://doi.org/10.1016/j.futures.2023.103236
Saage, R. (2007). Renaissance der Utopie. *UTOPIE kreativ* (201/202), 605–617.
Saage, R. (2013). New Man in Utopian and Transhumanist Perspective. *European Journal of Futures Research, 1*(1), 14. https://doi.org/10.1007/s40309-013-0014-5
Saage, R. (2016). Is the Classic Concept of Utopia Ready for the Future? In S. D. Chrostowska & J. D. Ingram (Eds.), *Political Uses of Utopia: New Marxist, Anarchist, and Radical Democratic Perspectives* (pp. 57–79). Columbia University Press.
Saage, R. (2019). Das Ende der politischen Utopie? In R. Saage (Ed.), *Das Ende der politischen Utopie?* (2nd ed., pp. 13–25). Suhrkamp.
Sand, M. (2016a). Responsibility and Visioneering—Opening Pandora's Box. *NanoEthics, 10*(1), 75–86. https://doi.org/10.1007/s11569-016-0252-7
Sand, M. (2016b). Technikvisionen als Gegenstand einer Ethik von Innovationsprozessen. In M. Maring (Ed.), *Zur Zukunft der Bereichsethiken* (pp. 333–354). KIT Scientific Publishing.
Sand, M. (2019). On "Not Having a Future." *Futures, 107*, 98–106. https://doi.org/10.1016/j.futures.2019.01.002
Sand, M., & Jongsma, K. R. (2016). Towards an Ageless Society: Assessing a Transhumanist Programme. In E. Domínguez-Rué & L. Nierling (Eds.), *Ageing and Technology* (1st ed., pp. 275–294). Transcript.
Sand, M., & Jongsma, K. R. (2022). AI in Medical Practice. In E. Di Nucci, L. Ji-Young, & I. A. Wagner (Eds.), *The Rowman & Littlefield Handbook of Bioethics* (pp. 328–339). Rowman & Littlefield.

Sargent, L. T. (2010). *Utopianism: A Very Short Introduction*. Oxford University Press.

Sargisson, L. (2012). Fool's Gold?: Utopianism in the Twenty-First Century. *Palgrave Macmillan*. https://doi.org/10.1057/9781137031075

Scheffler, S. (1993). *Human Morality* (Paperback ed.). Oxford University Press.

Segal, H. P. (2006). *Technology and Utopia*. American Historical Association.

Shew, A. (2022). The Minded Body in Technology and Disability. In S. Vallor (Ed.), *The Oxford Handbook of Philosophy of Technology* (pp. 516–534). Oxford University Press. https://doi.org/10.1093/oxfordhb/9780190851187.013.22

Thaler, M. (2022). No Other Planet: Utopian Visions for a Climate-Changed World. Cambridge University Press.

Turkle, S. (2011). *Alone Together: Why We Expect More from Technology and Less from Each Other*. Basic Books.

Tutton, R. (2022). The Sociology of Futurelessness. *Sociology, 57*(2), 438–453. https://doi.org/10.1177/00380385221122420

Valentini, L. (2012). Ideal vs. Non-ideal Theory: A Conceptual Map. *Philosophy Compass, 7*(9), 654–664.

van de Poel, I. (2021). Design for Value Change. *Ethics and Information Technology, 23*(1), 27–31. https://doi.org/10.1007/s10676-018-9461-9

Winner, L. (1986). Mythinformation. In *The Whale and the Reactor* (pp. 98–117). University of Chicago Press.

Wright, E. O. (2012). Transforming Capitalism Through Real Utopias. *American Sociological Review, 78*(1), 1–25. https://doi.org/10.1177/0003122412468882

Žižek, S. (2008). The Violence of the Liberal Utopia. *Distinktion: Journal of Social Theory, 9*(2), 9–25. https://doi.org/10.1080/1600910X.2008.9672962

Zohny, H. (2021). Future Versus Present Morality. In D. Edmonds (Ed.), *Future Morality*. Oxford University Press.

CHAPTER 4

How to and Where to Justice

Abstract Where ought we to interfere to create and ideally just society? Rawls' assumption that the primary locus of justice is the basic structure—comprising various socio-political institutions—initiates a discussion about the nexus of a just society. Technological utopias exceed the dichotomy between individualist and collectivist utopias and add technology as a third possible constituent for justice. They also challenge to rethink the background assumptions that underlie many theories of justice and are often seen as immutable. Doing this, advances our understanding of justice.

Keywords John Rawls · Basic structure · Constituents of justice · Scope of justice · Location of justice · Background assumptions

In previous chapters, I have discussed mainly criticism of utopianisms in general and technological utopias specifically. By way of an anti-anti-utopianism—an extended discussion of critiques of utopianism—I provided some reasons to assume that utopianism has a positive value. With an affirmative nod to different authors from Utopian Studies and political philosophy such as Ruth Levitas, Mathias Thaler, Frederic Jameson and Lucy Sargisson (see Chapter 2), I have suggested that utopias estrange us from the present and help us break free from the ideological imprisonments of our current views on the world and our

M. Sand, *Technological Utopianism and the Idea of Justice*,
https://doi.org/10.1007/978-3-031-75945-1_4

perceptions of the immutability of our current way of living and being. This estrangement might initially induce a cognitive change, plant a new idea or insight and might, thereby, become an important causal precursor to initiate actual change in the world. It must, however, not necessarily induce such change in order to be valuable and worthy of engagement. Utopias can help us see more clearly in which ways our current societies are defective, and they break the ideological spell that the present is unchangeable. They estrange by showing us, how the world could be different and better. The impact of such outlooks on individuals and decision-makers can be motivating, it can provide hope, where hope has withered. It can even galvanize action. While such outlook does not deliver direct guidelines on how to change the current socio-political system, this is a practically relevant function. It is practical in that understanding "what if" does help us to better identify in which way the present is deficient and move on from there. It is important to underscore that I do not claim that utopianism is the only device that can fulfill these functions. Art and literature are equally capable of causing a sense of estrangement and dissociation—they too can unsettle, induce discomfort and the feeling that things can be different and should be different.

In Chapter 2, I have outlined that utopianism could still carry value in the way just summarized, even if it were not directly relevant for governance or policymaking. There, I argued that estrangement and the subversiveness of utopianism are closely related, underpinning its cognitive function. The reader might still wonder at this point, how exactly *technological utopias* of the kind introduced above fulfill this function. In this chapter, I will further unpack how technological utopias estrange from and critique the present realities and concepts. The question, how subversiveness and critique are expressed in technological utopias *viz-a-viz* the notion of justice has been introduced via two different routes: The first route has been rather briefly introduced via Estlund's conviction that utopias are primarily about justifying principles of justice, which we have rejected as being too restrictive. Utopias might conduce to a better understanding of justice not by justifying principles, but by criticizing explicit and implicit background assumptions in prevailing theories. The second route has been opened by our discussion about the individual and her relationship and embeddedness in society as envisioned in the utopian tradition. Richard Saage's criticism suggested that utopias deserve enlistment in the canon only if society as a whole is tackled and imaginatively transformed by them. The primary focus on the individual in many technological utopias indicates their liberal capitalist roots, which they seem unwilling to discard. Authors—such as Saage—claim that the social obliviousness of technological utopias such as Transhumanism makes them non-utopian or worthless. We have seen that this presumes a non-reductive view on issues of the ideal society and justice.

These two routes ought to be synthesized and expanded upon in the presented chapter. First, I will argue that technological utopias urge us to question the background assumptions that many eminent theories of justice implicitly endorse—including, for instance, the idea that some forms of pluralism cannot be overcome by rational argument, that biological nature is immutable and so is the existence of limited altruism that result from it, and that some degree of material scarcity and informational limitations will prevail. Because of this, one could regard technological utopias as theoretical contributors to discussions about the scope and location of justice. In re-envisioning the interaction of individual behavior, society and technology to form new socio-technological constellations, they also re-envision how the theoretical boundaries between the basic structure as the primary object of justice and other normative realms blur. Technologies can be seen as antecedents for changing peoples' behavior and socio-political institutions. Technological utopias, in turn, re-envision how within such triangular constellations justice can emerge—underscoring technologies' indispensable role. This is done via challenging the scope and location of justice. These ideas are picking up on the strand of thought that was inaugurated at the end of Chapter 3.

That there is something like a scope to justice and a primary location of justice—with which technology possible interferes—is best understood through the idea of the basic structure set forth by John Rawls. The basic structure, according to Rawls, encompasses

> the way in which the major social institutions fit together into one system, and how they assign fundamental rights and duties and shape the division of advantages that arise through social cooperation. Thus, the political constitution, the legally recognized forms of property, and the organization of the economy, and the nature of the family, all belong to the basic structure. (Rawls & Kelly, 2001, p. 10)

Here, Rawls advances the idea that justice ensues primarily via socio-political institutions. What this does, crucially, is to separate the realm of justice from that of other normative realms such as that of morality and prudence. Samuel Scheffler suggests that "one of the most important tasks of the basic structure [according to Rawls] is to influence people's wants and aspirations, and another of its tasks is to regulate individual conduct in such a way as to preserve background justice" (Scheffler, 2010, p. 138). The motivation behind separating those two realms—the realm of the basic structure and that of individual morality—is driven by the insight that "rules of individual conduct are insufficient to preserve background justice [...]", as Samuel Scheffler suggests (p. 135). Upon reading this remark one might find in it right away a rejection of the idea discussed before that a just society could emerge purely based on individuals' compliance with moral rules and obligations—that, in other words, that there will never be a need to establish socio-political institutions that regulate the distribution of goods, assign rights and steer people's behavior.

Rawls' position suggests that he would radically oppose *technological utopias* that propose to foster justice and the ideal society merely by (technologically) transforming the individual. Aside from opposition to an apolitical—or better: non-institutional approach—to justice, Rawls' conception of the basic structure contrasts with *moral theories* that are all-encompassing—such as Utilitarianism—that suggest that there is one principle (e.g. the maximization principle) that ought to govern both individuals' conduct, the design of socio-political institutions and political decision-making. While the principle of justice that ought to regulate the design and instantiation of socio-political institutions shall have a bearing on the conduct of people and their inclinations, the rules that ought to govern individual conduct is not extensionally equivalent. The rules of morality could be much more comprehensive than the principles of justice are. This suggests that aside from principles of justice—such as the Difference Principle—moral theorizing could and should address "a wide variety of topics including supererogation, mutual aid, and personal virtues such as beneficence, courage, and mercy" (Scheffler, 2010, p. 130).

If we were to follow Rawls in assuming that rules of individual conduct are insufficient to preserve background justice, the question that emerges wonders reversely: Would the basic structure—a system of major socio-political institutions of a certain kind—be sufficient to ensure background justice? Since Rawls underscores the importance of individual morality as mentioned before, he is likely to deny a society would be ideal and just without moral duties and norms guiding individual behavior. But, even if not ideal, would such society be able to ensure some minimal form of justice—would it be able to ensure what Rawls' calls *background justice*, justice *by and large*? Cases of the following type seem to counter this assumption:

> Imagine a society S, whose constitution and legislation are committed to equality of opportunity between men and women, in that they prohibit official, explicit discrimination. Society S's educational system also tries to promote fair equality of opportunity through, among other things, an educational system that genuinely guarantees equal opportunities for access to women and men. However, the pattern of rules within the family generates strong pressure for women not to apply for competitive schools in the first place, and heavy (though informal) sanctions or those who decide to resist such pressure and apply. As a result of that, very few women attend competitive schools, receive good education, and engage in public life. Those who do, choose to do so at the cost of cutting their ties with their families and not finding a life partner. Given, for the sake of the argument, that fair equality of opportunity is the appropriate principle of justice, is society S just? (Ronzoni, 2008, p. 208)

Despite the perhaps well-justified "equality of opportunity"-principle and the respectively designed institutions, I reckon that most readers agree that the ensuing society S is not just. It might, thus, seem that without acknowledging the importance of moral rules that impact the "pattern of rules within the family", the institutions—no matter how great and well-defended they are—will not be just. This criticism is inspired by feminist critiques of Rawls and a subsequent critical discussion by G. A. Cohen, who suggests that institutions such as the ones in the example of society S leave "choices [...] open by those rules because neither enjoined nor forbidden by them" that undermine the justice conferring effects of said socio-political institutions (Cohen, 1997, p. 3).

However, we have already seen that according to Scheffler's interpretation of Rawls, posing those arguments against the idea of the basic structure is to misunderstand him and building up a strawman. In the above cited quote, Rawls explicitly acknowledges the "nature of the family" as a crucial element of the basic structure, and various other passages (emphasized in Ronzoni's analysis [2008]) suggest that he was not naïve about such circumstances and individual behavior that inhibit the justice conferring propensity of socio-political institutions. His approach does seem to cover cases such as these. The idea, as Scheffler summarizes in his analysis of Rawls, is "not to explain how individuals can be freed from the burdensome responsibility of securing social justice" (p. 135). Rather, one must assume that aside from the establishment of properly designed institutions additional moral considerations ought to bear on individual conduct, and that "individuals have a natural duty to support just arrangements, and that a just and stable society will be impossible unless the members of society have a strong and normally effective sense of justice" (p. 135).[1]

Whatever the correct interpretation of Rawls—whether he acknowledged adequately the necessity of addressing individual behavior with moral and institutional demands already in his original thoughts or will have to do so in order to respond to compelling objection by his critics—it seems plausible and realistic that both individual behavior (being it a result of habits, tradition or conscious choice—see footnote 2) and socio-political institutions mutually conduce to or potentially undermine the realization of a just society.[2] We can say that individual behavior impacts the effectiveness of social institutions in roughly the same way in which technologies affect social institutions, as suggested in the last chapter, and vice versa. (1) Individual behavior can undermine the effectiveness of social institutions (by exploiting loopholes, or as in cases of society S, by revoking antecedent conditions necessary to make them work), but

[1] This defense comes with theoretical costs, I believe: If the basic structure shall have a bearing on individuals' duties and responsibilities, as Scheffler suggests, why set the baseline of the basic structure, where Rawls' is setting it, namely accepting an inequality to incentivize productive people, who would otherwise not work as hard, rather than suggesting that the difference principle does not allow such inequality and *demand* that productive people work hard. In this light, Cohen's critique still looms large.

[2] Ronzoni's phrasing is quite intriguing, and it would be great to understand better what "pattern of rules within the family" refer to. For the sake of simplicity, I will identify such "patterns of rules" crudely as individual conduct. This is simplistic as it neglects whether these "pattern of rules" are better understood as traditions, culture or values of individuals. Depending on such reading, one might disagree about their relative mutability. In the triangle below (Fig. 4.3), this would impact whether we see them rather as exogenous (like justice itself a product of the complex interaction) to this triangle or endogenous (amenable such as institutions and social behavior). It is beyond the scope of the present work to discuss this in depth.

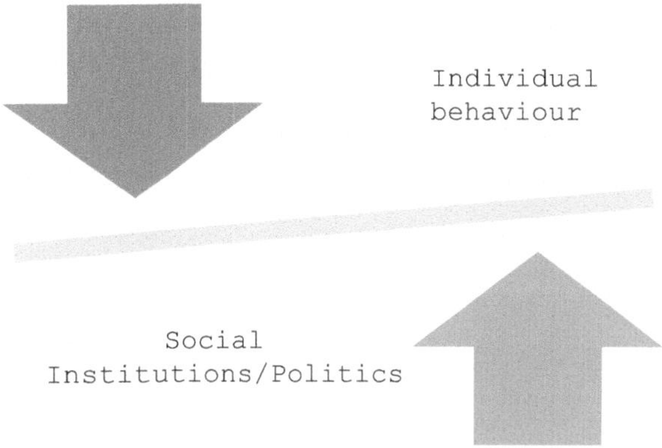

Fig. 4.1 Reciprocal relation between individual behavior and socio-political institutions

(2) also enhance the effectiveness of social institutions. Social institutions, in turn, can (3) undermine the effectiveness of good individual behavior (discouraging it) or 4) foster it (e.g. with incentives). The simplified image that emerges is, again, one of reciprocity (Fig. 4.1).

In order to establish a comprehensive picture of justice and its constitutive elements, we need to acknowledge another—third—relationship: Technologies and individual behavior impact another in similar ways reciprocally. This is a cornerstone of the postphenomenological philosophy that sees technologies as mediators of human's perception of the world and their interaction with it. Postphenomenologist Peter-Paul Verbeek suggests that due to their mediating quality, technologies become quasi-agents in the world and that they, in some sense, "answer" ethical questions: "Ethics is about the questions of 'how to act' and 'how to live' – and in our technological culture, these questions are not answered exclusively by human beings. By helping to shape the experiences and practices of human beings, technologies also provide answers to the central ethical questions, albeit in a material way. Artifacts are morally charged; they mediate moral decisions, shape moral subjects, and play an important role in moral agency" (Verbeek, 2011, p. 21). In my view, the idea that technology "answers" ethical questions actively can only be metaphorically understood (Sand, 2018). What the postphenomenological view, however, brings virtuously to the fore is that technologies

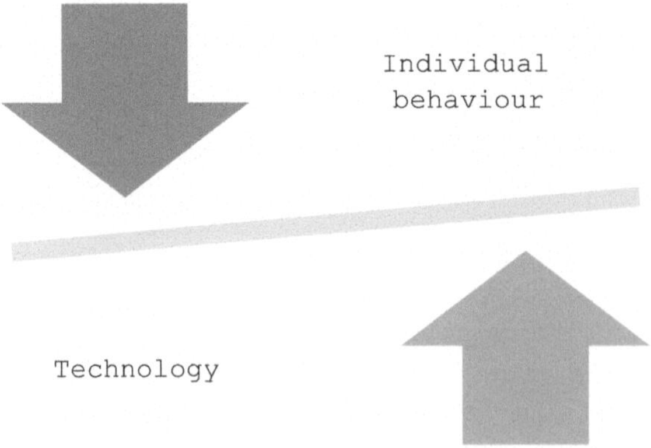

Fig. 4.2 Reciprocal relation between individual behavior and technology

can enhance or undermine the (moral) impacts of individual behavior. That seems plausible enough and—again—has a reciprocal character with roughly four possible directions. This can be exemplified with some rather mundane cases: (1) A digital calendar can help one remembering birthdays. Again, this does not imply seeing the calendar as an agent rather than as a mere tool. Before, we have spoken about the causal roles of social institutions without insinuating features of agency. (2) Social media might occupy your attention so much that you forget calling someone for her birthday. (3) A person reduces the impact of a technology by removing the technology—e.g. sunglasses when it gets dark. Or, (4) a person can employ technologies to enhance her activities' impact, for instance, by using a megaphone when speaking to a crowd. Each of these relations exemplify how technology and individual behavior reciprocally influence each other (Fig. 4.2).

If we take the three introduced reciprocal relationships together (Figs. 3.1, 4.1, and 4.2) and combine them to display how justice emerges through their various interactions with another, we arrive at the triangle (Fig. 4.3).

It is important to acknowledge that this picture, Fig. 4.3, portrays a realist and static imagine of the constituents of justice. Arguably, technologies' role as a pillar in this triangle is in and of itself the product of previous technological utopians' insistence that technology has a crucial role to play toward the emergence of justice. Still, there are three ways

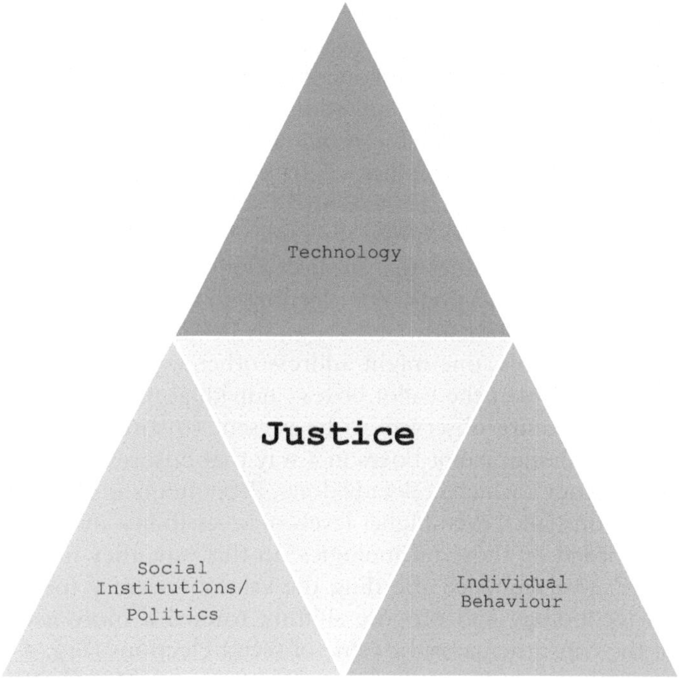

Fig. 4.3 Technology, social institutions/politics and individual behavior as constituents of justice

in which contemporary technological utopias continue to challenge this realist and static picture: First, by envisioning new constellations—relating to the order in terms of the sequence and symmetry of the pillars within the triangle and their interaction to instantiate justice, second by pressing the question, whether all of the pillars are necessary and third, by questioning the scope of justice that can emerge through such interaction. After briefly introducing the first two modes, I will finally introduce the idea of justice's scope as I understand it here as distinct from that of justice's location, which has been discussed *viz-a-viz* Rawls' idea of the basic structure.

The first way in which the speculations of technological utopians challenge, advance and remodel this image (Fig. 4.3) is by putting their finger on questions such as these: Where can and should individual obligations end and where will socio-political institutions and technology have to intervene and vice versa? This concerns—we recall again Rawls initial proposal—the question whether regarding particular instances of justice, any of the three pillars shall be the *primary* or *sole* focus of the intervention and to which degree each pillar will have to be addressed. One might say that the former concerns the *sequence* of a transformation process, the latter concerns the *symmetry* of the three pillars. This can be exemplified just as well with other values than justice: Consider the want for secret ballots to ensure free elections. For elections to be free, votes must be cast uncoerced and voluntarily, and they can be so only if they are cast secretly. To ensure this, one might address other voters or stakeholders to keep to themselves at the ballot boxes (individual morality). Additionally, one might require observers to be present (institutional solution). And, one might design ballot boxes in a way that ensures secrecy of vote casting (technology), which is already done. Proponents of digital tools to facilitate elections fancy even higher levels of accessibility and secrecy that might be afforded by those technologies. In this way, they insist to start the transformative sequence affording the value of secrecy for free elections with technology and propose shifting toward a more asymmetric picture of the constituents of the value of secret elections (Fig. 4.3).

We have advanced the rudiments of the general idea that in cases in which either socio-political institutions alone or individuals' duties alone are perceived as inadequate to generate justice—technology is introduced to intersect and *complement* these realms. The complementation with technology can also increase demands on individuals: In ongoing discussions about emerging digital health technologies, calls to redesign (educational) institutions and placing higher demands on *individuals* to ensure the just implementation of those technologies are voiced. Jongsma, Durán and I have argued that it will not suffice to redesign the digital technologies that are increasingly employed in health care to make them bullet proof against harmful outcomes, patient frustration and bias. The humans interacting with those systems will have to be educated and learn how to communicate and be attentive of these shortcomings (Sand et al., 2022). This requires institutionalization of education and regulation, as well as individual moral motivations of designers, manufacturers and other stakeholders involved. But, insisting on such interaction of the various pillars of justice also follows a practical concern (the governance of currently emerging technologies) and is, therefore, implicitly committed to some form of realism. Given the complex interaction, and dynamic shifting of responsibilities in socio-technical systems, addressing either only the institutions of the basic structure or only individuals will not suffice to create just health care—or more broadly speaking a just society. However, in these newly emerging and future socio-technical systems the boundaries between individual responsibilities and rights and socio-political institutions will continuously have to be redrawn. Underscoring this and interweaving it with a reading of Rawls' basic structure argument, Isaac Gabriel expresses poignantly how technology complements socio-political intuitions to form socio-technical systems that afford or undermine justice:

> [...] the basic structure of society is best understood **as a composite of sociotechnical systems**: that is, systems that are constituted through the interaction of human and technological elements. [...] **these elements interact dynamically to constitute new forms of stable institutional practice and behavior**. [...] AI interacts with the behavior of human decision-makers to shape the character of these practices, including how they distribute benefits and burdens across the population. (Gabriel, 2022, pp. 220 f.; own emphasis)

Gabriel's understanding of the basic structure, as expressed in the quote, is narrower than my own: We have seen that the basic structure might be the primary focus to ensure minimal or background justice, but it cannot constitute justice proper without additional moral constraints and demands directed at individuals. This also seems to have been affirmed by Rawls; an insight we have established through a critical reading of his account via Scheffler and Cohen. This is neglected in Gabriel's analysis. It should also be obvious that just like AI, on which Gabriel focuses here, other technologies (biotechnologies, nanotechnologies, etc.) can precipitate the emergence of new socio-technical constellations.

Of course, not only technologies have such impact: Whether or not we build the technologies that have this impact is shaped by the expectations that we associate with those technologies, which again are a product of the technological visions of them that have been established and circulated (Sand, 2018). My own view is, therefore, slightly different than that of Gabriel. My thesis is that technological utopias envision how technology could possibly interfere with and complement socio-political institutions and individual behavior and, hence, provide new views on where and how to interfere, if we want to enhance justice. Technological utopians establish technology as another pillar aside socio-political institutions and individual behavior toward the emergence of justice and interrogates how best to align them. Contra Gabriel, I also hold that the issue is not only the (envisioned) interference with the basic structure, but also the varying shifts between socio-political institutions and the other normative realms such as individual responsibility. We shall consider in the following chapter some examples how concretely technological utopias re-envision the alignment and order of those different pillars.

The second mode of challenging the realistic picture that technological utopians fancy is more radical. Aside from the question about the location of justice, which is what Rawls' basic structure analysis emphasizes and which concerns in our context the correct weighing of the different pillars in the triangle—utopians wonder whether technologies can entirely substitute either social institutions or the need for individuals to change their behavior to ensure justice. In this manner, technological means of intervening to equalize biological capacities promises to make forms of affirmative action redundant. We will get back to this example in Chapter 5. It should become clear now, why I deemed Fig. 4.3 to be a realistic representation rather than a utopian representation of the constituents of justice. Currently, these are the elements that undermine or advance justice. All three pillars interact in a complicated manner—each of them appears necessary to some degree. Some utopians force us to consider tilts to the symmetries that are sometimes alleged here. This realistic outlook, however, might be inacceptable from an ideal standpoint and does not invalidate the possibility that justice can evolve in very different manner, in the absence of one or two of its pillars, for instance. It is surely, a central mode of technological utopias to underscore the possibility of substituting changes to individual behavior and/or socio-political institutions entirely with new technologies to ensure justice.

Whether we can ever rid ourselves of the need of societal institutions that incentivize, coerce or regulate human behavior is, of course, a rather classic theme in the wider utopian tradition (Sargent, 2010, p. 91). Can it ever be the case that socio-political institutions—e.g. elections, parliaments, courts—will be expendable and will not be needed anymore to govern peoples' lives and their behavior to create a just society? G. A. Cohen's tractate on *Why not socialism?* (2009) uses the example of a camping trip to illustrate the principles of equality and community, which he believes to be desirable and foundational for socialism. On such a trip, resources and responsibilities are shared, and everyone works together for the common good without exploiting another. Cohen argues that this communal and cooperative ethos, which people find attractive in small, intimate settings, should extend to society writ large. He acknowledges the challenges in achieving socialism, especially in a world dominated by capitalist market structures, but contends that the moral case for socialism is compelling. Cohen does not offer a detailed blueprint for how socialism could be practically implemented but suggests that the desire for a more just and equitable society should motivate efforts to find ways to overcome these challenges (Cohen, 2009).[3] Crucially, no socio-political institutions seem required if the campers have the right inclinations toward another, which suggests that an ideally just society might be devoid of the need for political power and political institutions. The primary location of justice in this view is the ethos of the individual participant.

The particular question about the necessity of political power and political governance is traditionally in contention within the utopian literature (Saage, 2019). While we have seen through the insights of Douglas Mao—contra Saage—that most utopians acknowledge both socio-political institutions and individual behavior as being somehow constitutive for justice, they place different emphasis on those pillars. We have also underscored that some traditional utopias fall rather on an extreme side of the continuum between "individualist" and "collectivist": Just like some of these traditional utopias, such as Cohen's socialism that urge us to question whether all or only one of the pillars of justice that Fig. 4.3 outlines are needed, technological utopias press us to consider whether technology alone suffices to instantiate some forms of justice or justice writ large.

Lastly, as a third mode of challenging the realist image (Fig. 4.3), technological utopias also press us to rethink the *scope of justice*. We shall understand the question of scope as being concerned with the degree of mutability of the world. The scope of justice indicates the *richness* of justice in a society, how much flourishing and well-being it contains. The scope of justice suggests degrees of justness, with more or less flourishing, with more or less evil and misfortune in the world. A world in

[3] While Charles Mills and others complain that Rawls is too idealistic, Cohen suggests, thereby, that he is not idealistic enough in his insistence in the need of socio-political institutions.

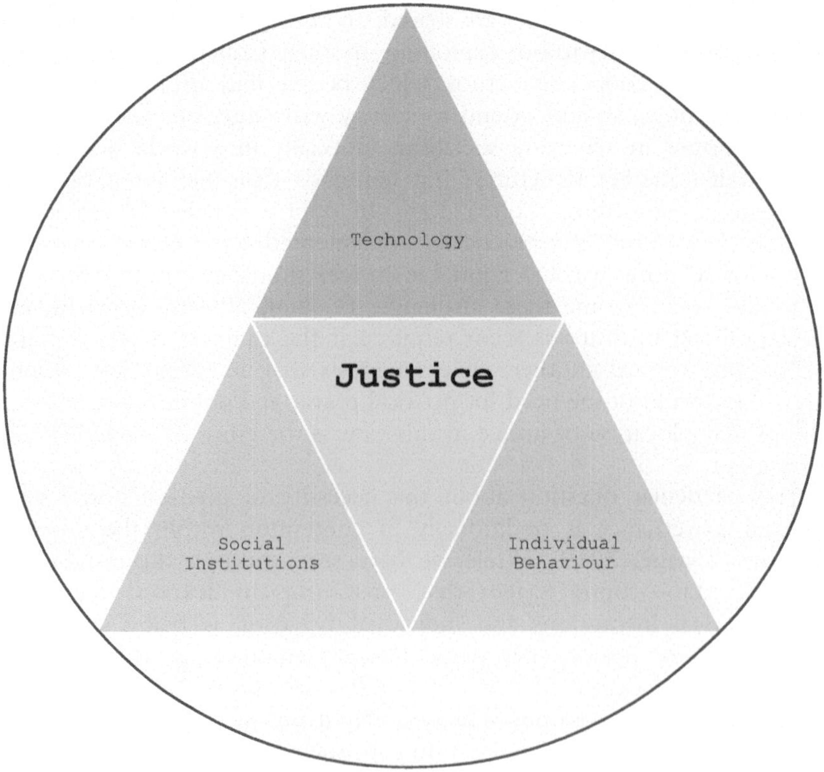

Fig. 4.4 Scope of justice

which human mortality prevails, is a world in which justice is achieved but perhaps to a lesser degree than in a world in which it has been eliminated.

In Chapter 1, I suggested that technological utopianism continuously presses us to reflect on alleged constraints—background assumptions—that are implicitly accepted in eminent justice theories and, thereby, increase our understanding of justice. These background assumptions in my understanding constrain how much flourishing and well-being can exist in the world—they constrain the scope of justice in other words. One can represent the idea of scope as a circle around the location of justice as expressed in Fig. 4.3. Differently wide circles could indicate the depth or richness of justice in a society (Fig. 4.4).

The idea of the scope of justice and its alleged background assumptions is best illustrated via a discussion of some ideas of Jacob Levy. In the first chapter and again above, we came across the idea that political and moral friction are the basis on which justice considerations commence. But agreeing with the fact that *some* moral friction is required to make justice considerations commence, does not yet say anything about *which kind of moral friction* we have to admit and see as immutable. In our discussion of Danaher's stagnation argument, we wondered whether and why we should renounce promises of a world of technological abundance and plenty, *assuming* that they lead to boredom and stagnation. We have seen that coercive stagnation might indicate a greater evil than voluntarily stagnation. Because the answers to those questions, about the desirability of abundance, plenty and the absence of any kind of friction are not obvious, as we have seen, such discussions might prove intriguing and undergird the typical cognitive and dispositional functions that undergird the positive value of utopianism: Such reflections could well induce the motivation to pursue overcoming constraints that have been unjustifiable regarded as immutable obstacles to a better life, but which they are not. What might those obstacles be? It is worth to quote Levy here in length, whose views we have already encountered:

> I mean to suggest that the Humean circumstances of justice, or something close to them, are not waivable assumptions in theorizing about justice; they help to set the meaning of the inquiry. [...] Mutual vulnerability (Hobbes), mortality (Heidegger), natality and plurality (Arendt), limited knowledge and information (Hayek, Oakeshott), the circumstances of politics (Waldron), linguistic communication rather than telepathy — **these are the kinds of considerations that enter in at the beginning of theorizing the human, social, political condition**. Perhaps some of them are more fundamental than others; perhaps some are not fundamental at all. [...] But it is with good reason that the great theorists so often began with their various accounts of human limitations, and a theory that is constructed without reference to *any* of these is neither a **theory of justice** nor **a theory of politics** but a branch of speculative fiction. (Levy, 2016, p. 332; own emphasis)

Levy suggests that the earlier discussed example of the camping trip that Cohen uses to introduce the principles of ethos and community as foundations for a just society, is one such example of theorizing that is closer to speculative fiction than to theories of justice (p. 319). Interestingly, neither of the unwaivable assumptions that Levy lists here are relinquished in Cohen's vision. Limited altruism might be a condition that Cohen indeed eliminated from his imagined society: But, how unwaivable is *that* condition? More generally, this leads one to wonder: Is the list exhaustive, is it complete? Do all items have to necessarily be on it? Initially, one can agree that theories of justice must commence with *some* background assumptions, but that does not yet tell us *which ones* and how exactly do they weigh in on a theory of justice—*to which degree do they have to be taken into account*? Do any of those mentioned by Levy belong to the list of those we have to commence our theory with? As mentioned before, I doubt that metaphysical immortality can be achieved. Mortality seems to me to be an inevitable aspect of human life. Crucially, however, the problem of justice is usually not located in this fact of inevitable mortality, but rather in the observation that life expectancy on a global scale is for various reasons unequally distributed. Some people get to know their grand-grandchildren, others do not even reach adulthood. To which degree this could and should be changed, is much more contested in the literature on justice than the basic fact of human mortality: What are the biological or broadly speaking natural causes of unequal life expectancy, what are the political causes and how do they intersect and where can we interfere? If biological and other natural causes that affect life-expectancy could be amended—which is undeniable already the case to some degree through the comprehensive medical interventions that characterize modern societies and other engineering feats that protect us from natural catastrophes and suggested by some technological utopians to become even more comprehensively mutable in the near future—then how much effort shall we put into this cause? Are increases in life expectancy or increases of quality of life the better targets for health policy? That is typically the derivative questions that find their way into more applied theories of justice, when human mortality is taken serious as a background assumption (Hughes, 2004). Equally, Levy assumes the background assumption of informational limitations rather lightly and swiftly. To which degree do we have to consider this as a fact and in which way does it weigh in on theories of justice? How limited is our knowledge really and of what subject is knowledge limited? Of course, there are limitations to our knowledge about the intentions of other agents, which gives rise to dilemmatic situations concerning global security. Then, there are epistemic limitations about what consequences diverse political interventions bring about (a point that Popper made abundantly clear, and which led him to suggest a more piecemeal approach). However, looking at this from such general level, these limitations of knowledge and information are of very different kind and concern very different objects. Again, it is unclear whether we have to accept all of these facets of informational and knowledge limitations to weigh in on theories of justice and, if so, how. The utopian discourse itself could conduce to answering these questions, I believe, and illuminate these nuances.

To summarize, Levy has not defended any of the background assumptions that are listed exemplarily. Even *if* we were to assume that a theory of justice must take *some* of these considerations into account, we have seen that it is unclear whether these or others should be taken into account and, if so, how. How fundamental are knowledge limitations and what exactly is the problem of mortality, if it is one that ought to be considered by justice theorists? However, do we even have to assume that any of these considerations have to be taken into account? To me, the relentless and insistent posing of that question underlies the utopian spirit and is the source for earlier innovations in our political thinking and the dismantling of previously held beliefs about immutable aspects of the way we live. Before moving on to a discussion of some examples of technological utopias probing background assumptions and the location of justice, it will be worth considering another one of Levy's assumptions: It seems that this viewpoint is established with appeal to a rather narrow conception of the purpose of political theories and theories of justice. The following longer quote neatly interweaves questions of justice that concern us here with our overall topic of technological utopias as speculative works of fiction:

> The principles appropriate to political and legal life are partly constituted by the problems and limitations of human social life; they are not imported from the realm of moral truth and then applied to a more or less recalcitrant world. [...] I quite like speculative fiction, and have spent more than one late night discussing a topic like 'how would society be organized if we had technology that succeeded in defeating the problem of scarcity, or immortality, or a Foundation-style solution to information problems?' This kind of thing is good fun and can sometimes cast interesting light on some feature of social organization. That does not make it something we must engage in, or even usually should engage in, as a first step to understanding the question of practical reason, what is to be done in the world we inhabit. (Levy, 2016, p. 333)

The passage raises another range of intriguing questions: What are those "interesting features of social organization" that fiction could reveal and—directly following up on this—why are those not at least tangentially interesting to justice concerns? I agree with Levy and have done so in previous sections of this book that "what is to be done in the world we inhabit [now]"—at least in the narrow sense of governing society—cannot and probably should not be directly answered by looking at and engaging with the technological utopias that have been presented and discussed in the present essay. However, I have underscored that there are other cognitive functions that those narratives can fulfill. Underlying Levy's reasoning is the idea already familiar to us from the previous discussion that political theorizing and theorizing about justice needs to pursue a practical goal, namely answer the question "What is to be done in the world?". However, how that practical goal ought to be achieved remains entirely unclear and there are certainly two import points to consider.

First: What is required from a theory of justice in order to fulfill that practical goal, *if* we were to endorse it for the sake of argument? Ought it to answer for each and every actor what they ought to do? Then clearly most if not all political theories and theories of justice have failed to some degree: Most political theories provide some higher-level reflection—e.g. arguing for the rights of refugees—without being very detailed about who should ensure and legislate such rights and how such rights can be guaranteed by the highly diverse political actors across the globe that are dealing with migration. Few, if any, of such more realistic proposals outline in much detail why this or that political party should coalesce with this or that rival party and then write this or that item into their coalition agreement and instantiate it with the support of this or that stakeholder. And even those suggestions are in a sense higher level than the fine-grained realities of practical politics, where much diplomacy commences by determining who sits next to whom at the table.[4]

Second: As the notion of practical relevance is ambiguous—we are not forced to understand it as merely referring to changes in terms of institutional politics and governance, etc. As we have discussed in Chapter 2, utopian narratives can be enticing, inspiring and motivating—they can induce hope that things might change, which is practically relevant and, of course, they also have more concrete impacts on more applied debates

[4] I have in mind the highly readable contributions to Deen Chatterjee's *The Ethics of Assistance—Morality and the Distant Needy* (2004).

in technological decision-making: Evidence for the fact that such narratives have a role to play in more concrete debates about technological governance and socio-political debates and that they play a role in actual technological and engineering endeavors is abundant (see, for instance, McCray, 2013). Unlike the main addressee of Levy's argument, I do not claim that there is a *primacy* of the type of reasoning induced by utopianism that should come before more realistic explorations. This is the central, motivating concern and the target that Levy wants to refute—that political philosophy should be primarily and perhaps exclusively be concerned with ideal theorizing (utopianism) (Levy, 2016, p. 316).[5] While I do not defend the primacy of engaging with utopianism to understand justice and create a more just society, I am convinced that at some point—given the distribution of cognitive labor that already characterizes the social sciences and humanities—we can and should allow ourselves to ponder what it would mean, if a few of the aforementioned background assumptions would be relinquished and to think about what it would require to relinquish them. But, again, is this a fruitful exercise? Is it an exercise in better understanding justice—or is it an exercise in something else? I do not know what justice is and what it would require, but it strikes me as odd to accept that justice is what we can realize given these constraints, and not *also* an endeavor of questioning whether we can overcome these constraints. In the absence of valid proofs of the inevitability of such realistic assumptions that theories of justice need to take into account—which none of the authors who are affirmatively cited by Levy have provided to my knowledge—one can sustain the conviction that it is a central strength of utopianism to persistently reopen the investigation into their mutability. Technological utopias such as the afore discussed narratives provoke reconsideration of what assumptions we take for granted when we discuss and theorize in which ways the world would have to change to allow for a better way of living and being and whether all constituents currently deemed necessary to achieve justice (individual behavior, technology socio-political institutions) need, in fact, be addressed. This shall be expanded upon in the next chapter.

[5] He cites this passage from *A Theory of Justice*: "The reason for beginning with ideal theory is that it provides, I believe, *the only basis* for the systematic grasp of these more pressing problems. At least, *I shall assume that insight can be gained in no other way*, and that the nature and aims of a perfectly just society is the *fundamental* part of the theory of justice" (Rawls 1999, p. 8).

References

Chatterjee, D. K. (Ed.). (2004). *The Ethics of Assistance—Morality and the Distant Needy*. Cambridge University Press.

Cohen, G. A. (1997). Where the Action Is: On the Site of Distributive Justice. *Philosophy & Public Affairs, 26*(1), 3–30. https://doi.org/10.1111/j.1088-4963.1997.tb00048.x

Cohen, G. A. (2009). *Why Not Socialism?* Princeton University Press.

Gabriel, I. (2022). Toward a Theory of Justice for Artificial Intelligence. *Daedalus, 151*(2), 218–231. https://doi.org/10.1162/daed_a_01911

Hughes, J. (2004). *Citizen Cyborg: Why Democratic Societies Must Respond to the Redesigned Human of the Future*. Westview Press.

Levy, J. T. (2016). There Is No Such Thing as Ideal Theory. *Social Philosophy and Policy, 33*(1–2), 312–333. https://doi.org/10.1017/S026505251600025X

McCray, P. (2013). *The Visioneers: How a Group of Elite Scientists Pursued Space Colonies, Nanotechnologies, and a Limitless Future*. Princeton University Press.

Rawls, J. (1999). *A Theory of Justice* (Revised ed.). Harvard University Press (Belknap).

Rawls, J., & Kelly, E. (2001). *Justice as Fairness: A Restatement*. Belknap Press of Harvard University Press.

Ronzoni, M. (2008). What Makes a Basic Structure Just? *Res Publica, 14*(3), 203–218. https://doi.org/10.1007/s11158-008-9056-0

Saage, R. (2019). Gibt es einen anarchistischen Diskurs in der klassischen Utopietradition? In R. Saage (Ed.), *Das Ende der politischen Utopie?* (2nd ed., pp. 26–45). Suhrkamp.

Sand, M. (2018). *Futures, Visions, and Responsibility: An Ethics of Innovation*. VS Verlag für Sozialwissenschaften.

Sand, M., Durán, J. M., & Jongsma, K. R. (2022). Responsibility Beyond Design: Physicians' Requirements for Ethical Medical AI. *Bioethics, 36*(2), 162–169. https://doi.org/10.1111/bioe.12887

Sargent, L. T. (2010). *Utopianism: A Very Short Introduction*. Oxford University Press.

Scheffler, S. (2010). Is the Basic Structure Basic? In *Equality and Tradition—Questions of Value in Moral and Political Theory* (pp. 129–159). Oxford University Press.

Verbeek, P.-P. (2011). *Moralizing Technology: Understanding and Designing the Morality of Things*. University of Chicago Press.

CHAPTER 5

Justice in Technological Utopia

Abstract In this chapter, it will be outlined in more detail and with some examples, how technological utopias relocate justice from institutions and individuals, for example, to human biology. The idea is controversially discussed and need not be endorsed to fulfill its illustrative function for the purpose of the present argumentation. Additionally, technological utopias suggest the possibility of reconciling incompatible views of the good life in virtual reality and resolving the problem of scarcity that underlies some of the most prominent theories of justice.

Keywords Scarcity · Human nature · Eugenics · Reconciling pluralism · Meta-utopia · The good life

It is worth pointing out that my defense of technological utopianism as tools that help us to re-envision the relationship between the individual, the social and the technological and, therefore, advance our understanding of the scope and location of justice is not—as mentioned before—a vindication of the category of technological utopianism. It salvages only those technological utopias that carry this feature. At this point, I clearly owe a more in-depth discussion, in which way I consider the technological utopias presented in the first chapter achieving this task.

M. Sand, *Technological Utopianism and the Idea of Justice*,
https://doi.org/10.1007/978-3-031-75945-1_5

5.1 Immutable Human Biology and Its Impact on Talent and Equality

This first dimension challenges underlying assumptions about what factors of the world are mutable, appertaining in particular to our biological constitution. Many theories of justice (e.g. Rawls) take as a baseline that we are exposed to some sort of natural lottery, which causes differences in abilities and talents, which again affects the extent to which we can flourish. Political institutions then ought to outbalance those effects. These theorists shy away from addressing those biological roots, because they consider them unchangeable and in doing so, they purport ways of creating justice that take those as a given. When it comes to grave unluck regarding someone's health—Thomas Nagel believes that societies have great duties to nullify the unfairness that arises because of the natural lottery: Social institutions must be designed so as to enable people with illnesses or disabilities to fulfill their basic needs and lead a decent life (p. 107). But the real problem emerges according to Nagel in a different domain when it comes to the distribution of benefits and advantages based on talent and effort. He suggests that many egalitarians wish to grant rewards that are achieved through personal effort (politically and economically). If two people A and B had the same equality of opportunity to achieve X (say, win a sprint), and B, due to her efforts succeeds in achieving X, she deserves the advantages that are tied to X (say, prize money). But such products of effort, Nagel contents—even though conceptually distinguishable from talent—are in practice so tightly interwoven with natural talent (and, in fact, class) that they reciprocally impact another. Since, talents are naturally endowed and cannot be changed, societies striving for equality and justice will simply fail to be able to realize the ideal of a society that endows people with benefits purely based on their efforts:

> [I]t is impossible in practice to disentangle the effects of talent [which are natural and involuntary] from the effects of effort, since effort is expended through the exercise of talent, and talent develops into a valuable ability only through effort. [...] So if one does not object to inequalities due to effort, reluctance to prevent them will automatically carry over to the effects of talent that go with them. To be sure, effort also combines inextricably with class in the causation of inequality: Those with privileged background and education, not to mention money, can profit more from

> a given sate for that through measures of positive equality of opportunity, **whereas nothing can be done to equalize natural abilities**. (Nagel, 1991, p. 119; own emphasis)

Transhumanists disagree. At least, they desire differently. When Nagel suggests that there is no justice problem in the direct results of the natural lottery, but only in the indirect consequences regarding how society apportions benefits and advantages based on those natural differences (Nagel, 1991, p. 107), which are arguably mutable, the Transhumanist intervenes and suggests that that scope of social influence on nature could—and should—expand. Transhumanists shift the accentuation of the location of justice regarding this question away from both individual responsibility and social institutions—both of which Nagel deems incapable of separating between what the justified sources of reward (effort) and those that are not (talent)—toward technology. When it comes to talent and effort, such separation might in practice be technologically equally impossible. Still, other causes that hamper the flourishing of one's life—illnesses and disabilities—have more easily identifiable biological causes. James Hughes proposes to utilize modern technology to nullify nature's impact on the differential flourishing of people: "A political movement based on both technological progress and individual liberty will then see ways that democratically regulated and distributed, freely exercised technology can create a more equal, empowered and united world. One way is by reducing the **biological bases of social inequality**" (Hughes, 2004, p. 195; own emphasis). Thus, Transhumanism emphasizes the mutability of our genetic makeup, which they see as an underlying cause of many inequalities. In relation to the sketched image—rather than looking at the institutional side or individual responsibilities, they emphasize technologies' potential, thereby shifting the focus of the question how justice could be established. Critics suggest that the view is a product of the exclusivity of the movement (white, male, able-bodied) and their respective biases (Chapter 3): Transhumanists perpetuate a utopia in which disability (and other forms of diversity) are eliminated rather than accommodated (Shew, 2020). An anti-ableist utopia, in contrast, could conjure various ways in which diverse bodies and minds are enabled to live flourishing lives together, rather than being eradicated, focusing on cultural or individual change or socio-political institutions as a locus (Smith, 2021). Of course, such utopias could also fancy technology as the primary nexus for establishing a just, anti-ableist society. Juxtaposing these two very different technological utopias, appears to be a highly intriguing undertaking for future research. The methodological nature of the present study generates no need to commit to either of these approaches, as I will underscore repeatedly in the following.

There are many skeptics who doubt the degree to which genetic interventions will enable modification of certain traits: Adam Rutherford suggests that many differences in abilities (e.g. levels of intelligence) and skills as well as many diseases (e.g. alcoholism and other mental health issues) have no unanimous genetic cause. Hence, no interventions will ever prove to become a sliver bullet (Rutherford, 2023). Such criticism of the underlying scientific assumptions of genetic interventions is persuasive. Still, if differences in talents and abilities continue to contribute to large degrees to differences in flourishing in life, and society has proven incapable of outbalancing those differences in the past decades, must one really resign and accept them? Should we pursue the elimination of the biological causes of these differences or rather accommodate them by providing enabling environments? Nick Bostrom suggests in the controversially discussed article that was cited before:

> All of this is based on the hypothesis that germ-line engineering would in fact increase inequalities if left unregulated and no countermeasures were taken. That hypothesis might be false. In particular, it might turn out to be technologically easier to cure gross genetic defects than to enhance an already healthy genetic constitution. We currently know much more about many specific inheritable diseases, some of which are due to single gene defects, than we do about the genetic basis of talents and desirable qualities such as intelligence and longevity, which in all likelihood are encoded in complex constellations of multiple genes. If this turns out to be the case, then the trajectory of human genetic enhancement may be one in which the first thing to happen is that the lot of the genetically worst-off is radically improved, through the elimination of diseases such as Tay Sachs, Lesch-Nyhan, Downs Syndrome, and early-onset Alzheimer's disease. This would have a major leveling effect on inequalities, not primarily in the monetary sense, but with respect to the even more fundamental parameters of basic opportunities and quality of life. (Bostrom, 2003, p. 503)

This paragraph from *Human Genetic Enhancements: A Transhumanist Perspective* is instructive in various ways. It suggests that making better people, does not necessarily commence first by making those better that are already well-off. Hence, it is not necessarily an anti-egalitarian view. To the contrary, the view presented here is prioritarian in that it addresses those that are currently (in Bostrom's view) worse-off than others. Of course, Bostrom arrives at this proto prioritarian view by pondering the possibilities of current genetic technologies and rather than by evoking political ideals. Still, the passage might exemplify once more—contra Saage—that the *object of change* of this technological intervention into human biology can be society as a whole, while *the agent of change* is technology as applied to the individual (recall Chapter 3). Bostrom fancies technological interventions as applied to individuals, while considering the relationship *between* individuals and their comparative standing in terms of well-being. The view is concerned with justice, and it locates this concern at the level of human biology with the possibility of technological interference. The view suggests that the general enhancement project with its goal of making us collectively better-off, can plausibly commence through first leveling the worst-off up by eliminating biological sources of misery and suffering. In this vein, many Transhumanists, including James Hughes, display collectivist and even egalitarian sentiments regarding access to technological tools and the goals of their application. The above quote suggests that the technological utopia of Transhumanism challenges the view that justice is primarily located at the level of socio-political institutions that distribute rights and goods in a society. It arises out of the perceived deficiency of relying on redesigning socio-political institutions to create a just society. It, thereby, challenges and provokes us to rethink a key aspect of many peoples' ideological or mental imprisonment; namely that misery and suffering are and must remain a necessary part of human existence.

Clearly, as we already touched upon the anti-ableist critique, this is a contentious matter: Bostrom suggests in the above quote that a person with Down Syndrome counts among the "genetically worst-off"—a judgment that many will find appalling and one that can be easily disproven with examples of individuals flourishing with this condition. The history of eugenics embarks its horrific journey based on the notion that some lives are more worthy of living than others and we find such convictions clearly in those remarks by Bostrom. It is natural to wonder: Who does he think he is to declare which lifes are worth living and which are not and does not the rhetoric itself conduce to stigmatization and social exclusion of those individuals? Is the haunting specter of eugenics and its horrific manifestations not in and of itself a reason to not even tinker with the idea of expanding the notion of justice into the realm of human biology as Transhumanists are eager to do?

But, sure enough—while the idea of forceful and coercive population control is horrific and perhaps even the liberal version of this idea is problematic in its perpetuation of stigma and the naïve belief that parents' decision which kind of children they get would be ever completely voluntary—it is at first sight no less problematic to assume the contraposition, namely that societies have to bear the burden of whatever its future offspring will be and merely try their best to alleviate the accompanying suffering (Kitcher, 1996, p. 198). First, thus, this demands that one also puts into question the moral acceptability of a *laissez faire* approach to population control. Questioning and critiquing the *status quo*, as we have suggested before, is an integral part of utopian speculation. Second, it seems misguided to morally equate all forms of population control: Paying someone to not get pregnant, carries *prima facie* a very different moral quality than sterilizing someone forcefully.[1] Philip Kitcher writes in an agreeable manner that "[…] it is overwhelmingly obvious that some varieties [of eugenic practice] are far worse than others" (1996, p. 193). Given the ferocious atrocities committed by the Nazis, Bostrom himself suggests that "we ought to think twice before giving our support to any proposal that would have the state regulate what sort of children people are allowed to have and the methods that may be used to conceive them" (Bostrom, 2003, p. 499).[2]

[1] Adam Rutherford (2023) does not provide a concise definition of eugenics. He mentions these examples side by side suggesting a moral equivalence.

[2] In texts of the Transhumanist forerunner Julian Huxley, we find overwhelming evidence that he was a proponent of strong eugenic measures including coercive sterilizations (Weindling, 2012). When it comes to the scope of his envisioned interventions, he rejects racism for empirical and for normative reason: "[…] investigation has conclusively demonstrated first that there is no such thing as 'pure race'. Secondly, that the obvious differences I level of achievement between different peoples and ethnic groups are primarily cultural, due to differences not in genetic equipment but in historical and environmental opportunity" (1992, pp. 256 f.). In another passage, in the same article, he suggests regarding "haemophilia, colour-blindness, mongolism […]" that these are "severe and primarily genetic disabilities". Here, too, we find a quality-of-life judgment not only of "mongolism", but oddly also of color-blindness. Huxley suggests: "All new dominant types begin their career in a crude and imperfect form, which then needs drastic polishing and improvement before it can reveal its full potentialities and achieve full evolutionary success. Man is no exception to this rule. He is not merely exceedingly young; he is also

Thirdly, rather than delving deeper into whether or not the Transhumanists' suggestions point at a proto justification of some forms of eugenics and the subsequent concern whether all of these are to an equal degree morally objectionable, I shall emphasize again which role the example plays in my analysis.[3] My own stance is a methodological one: Suggesting that it is worthwhile to ponder whether technology would be the ideal means to support the creation of a just society via biological interventions does not entail a commitment to that particular utopia. In other words, while I believe it is worthwhile to think about human biology as a potential locus for justice, I am not forced to accept that this is in fact the right perspective, let alone that it should be striven for. The example serves a much more limited purpose.

5.2 Material Scarcity—Or How Flourishing Can a Society Become?

One could interpret Transhumanism as insinuating a divergence from the question of justice as "What is everyone's due?" to "What if everyone could get what they wanted without depriving others?" One could think about it in this way: Transhumanists might question that there is anything wrong with some having a lot more than others, if those others also have a lot. Earlier, we have introduced Amartya Sen's thought-experiment,

exceedingly imperfect, an unfinished and often botched product of evolutionary improvisation" (pp. 253 f.). That this does not include uniformity but rather variability in genetic makeup is later also expressed. What is also expressed is the conviction that patients with transmittable gene defects shall be "persuaded to not reproduce" (p. 268). Further, while he is initially skeptic about genetic origins of many mental diseases, he writes that "the so-called social problem group", by which he means the people "who seem to have ceased to care, and just carry on the business of bare existence [...]". They have to be supported "out of public funds and become a burden on the community. Unfortunately, they are not deterred by the conditions of existence from carrying on with the business of reproduction [...] Here again, voluntary sterilization could be useful. [...] Compulsory or semi-compulsory vaccination, inoculation and isolation are used in respect of many public health risks: I see no reason why similar measures should not be used in respect of this grave problem [...]" (p. 270). Even within their historical context, these expressions are detestable. The stigma that poverty is a cause of the individuals' underlying biological defects rather exacerbates than tackles an existing injustice. Such sentiments taken together are troubling to say the least and they continue to fuel skepticism about the morality of Transhumanists' envisioned biological interventions.

[3] A valuable investigation into the wider relationship between utopianism and population control has been provided by Parrinder (1997).

where he proposes that a flute must be distributed to one of three children. Sen uses the case to explicate the idea of genuine pluralism that cannot be resolved by way of transcendental reasoning and wants to unravel the epistemic absolutism underlying ideal theorizing in political philosophy (Sen, 2010). We recall that one of the children has the most talent to play and would, therefore, provide most joy to the listeners—she *creates most pleasure*. Another one, while not being as talented and capable, has built the flute. In other words, she would ***deserve*** it most. Lastly, there is a child, who is overall the most miserable in terms of her well-being: She *needs* it most, although she is neither talented nor did she build it. While utilitarians might favor the child, who causes most pleasure, libertarians might argue that the one who built it deserves it. Sen, in contrast, suggests that there is no transcendental style approach that can resolve the case satisfactorily and unequivocally.

There are two essential presumptions in this though-experiment that causes the predicament: Scarcity as the availability of only one flute and the assumption that once one of the children has the flute, the others cannot access and use it anymore. Once handed over to a child, she is somehow deemed as the "owner" of the flute, which means restricted or no access for others. It is precisely those kinds of presumptions that utopians—including the technological utopians we encountered before—challenge: Why not delivering three flutes or arrange the "possession rights" in a way so that each of the children can play with it one after another? Technology can help establishing those arrangements: Consider, for instance, stopwatches to determine the timeframes of usage as a simple example. That would resolve the fundamental dispute that Sen identifies, before it even emerges. Michael Hauskeller suggests in various articles that Transhumanism is—like the utopia of Cockayne—a utopia of abundance. It envisions a world in which (material) desires can be maximally fulfilled (Hauskeller, 2012). What Transhumanists ask us (among a lot of other things) is, thus: Must we really accept that for many people to have a decent quality of life, with sufficient food, shelter and education to provide some basic opportunities of flourishing, that others must give up some of their goods? Does widespread flourishing require that some peoples' liberties, their freedom to own things, must be curtailed? Could there not be plenty for everyone? A view that supports the creation of plenty and widespread distribution of various material goods for everyone, does not necessarily entail that everyone needs to receive the same: There could remain varying degrees of plentitude. In this way, we see that

Transhumanism is at least *prima facie* compatible with very different substantial views about justice (sufficientarianism, prioritarianism, egalitarianism). One might think that the question about the scope (Which background assumptions of justice are immutable?) and location of justice (What is the primary source or location of justice?) are distinct from the substantive question about justice (What is justice?). In short, one might think that one can endorse, for instance, Rawls' view on the basic structure and be a libertarian, a sufficientarian or limitarian. That might be true only to some degree, however: Libertarians, for instance, might argue that the efforts of building the flute that one of the children invested, makes her deserve the flute. While they might acknowledge the values of need and pleasure, they might claim that it would constitute an *unfairness* to her, if the others also got flutes without having invested any effort. Furthermore, they might claim that distributing flutes also to the other children will disincentivize her in the future from honing her crafts and abilities. Conclusively, they might regard it as bad, if all children in this example had access to flutes. In short, concerns about the location and the scope of justice are in a complex way intertwined with substantive questions about justice. As sketched, a desert-based libertarian view might suggest that there should overall be limited amount of goods to prevent the unfairness that the undeserving have the same as the deserving. We cannot delineate these interrelations regarding the present examples in detail—that would bring us too far off the track. However, the existence of these interdependencies warrants the conclusion that advancing our understanding of the scope and location of justice also increases our understanding of a more substantive idea of justice.

Most often the desirability of some form of abundance is associated with food. In the myth of Cockayne, food features prominently. The land is displayed as a place where "roasted pigs wander about with knives in their backs to make carving easy, where grilled geese fly directly into one's mouth, where cooked fish jump out of the water and land at one's feet" (Pleij & Webb, 2001, pp. 89–106). The availability and distribution of food has long played a role in utopian thinking and recent technological advancements continue to fuel fancy and desire in this regard: Rather than having pigs wander about with knives in their backs, these pigs are envisioned as growing out of petri dish (Castle, 2022). Other goods, however, are of a different kind and one must wonder whether an abundance of them is as desirable as an abundance of food.

Empirical studies suggest that some goods are attributed with a relative value. The desire to own them seems to depend to some degree on their distribution among other people. The perceived good of houses of a particular size, for instance, seems to depend on how big houses in their vicinity are (Frank, 2011). Regarding these goods, it would obviously be mute to have more or bigger ones of them. The availability of more and bigger houses would merely engender the desire in others to have even bigger houses than those existing. Whether this psychological mechanism can be saturated at one point is unclear. As the widespread availability of some goods merely contribute to the desire for more of those goods—or, in fact, the possession of other goods of a similar kind—the fear that the promotion of an infinite abundance of goods that are mere luxuries (such as big houses) leads to further environmental degradation looms large. We have seen before that the rhetoric of abundance of some Transhumanists and other contemporary technological utopians awakes associations of classic capitalist rallying cries. Critics detect in these notions not a concern for the disposition of certain goods in furthering quality of life, but merely for their ubiquitous availability. They see worshippers of a hollow materialism (see Chapter 3). Promoters of a degrowth utopia juxtapose these visions of technological utopians with their own, suggesting that we must renounce the paradigm of growth and establish socio-political institutions that help us maintaining, repairing and sharing existing technologies and goods rather than producing more and new ones (Kallis & March, 2015). However, does all striving for material abundance have to have these consequences? Could one not have ways of partaking in certain goods without the degrading environmental effects that the past decades of industrial production engendered?

Lawrence Lessig has early argued that via digital technologies some goods—like works of art, film and music—could be made available to a much larger audience for much lower prices, which is a desirable thing (Lessig, 1999). Whereas stealing a CD some years ago meant that a potential buyer was deprived of the opportunity to listening to such piece of art, the digital realm enables copying files infinitely so that everyone can eventually participate. Hence, the digital realm seems to hold the promise that enjoying some goods could be made possible without having to materially reproduce them.[4] In this way, the possibility of abundance

[4] I mentioned before that this fancy has also been criticized. Digitization's "hidden" environmental impacts are increasingly underscored (van Wynsberghe, 2021).

also features prominently in the technological utopia of virtual reality. Why worry about owning a yacht and flying on holidays, with all its detrimental environmental impacts, if such experiences can be reproduced in a virtual reality? In the following passage, David Chalmers ties various streams of our previous discussion together:

> In the short term, while virtual worlds are inferior to the nonvirtual world, virtual abundance may have at most a small impact on our lives. But if virtual realism is correct, life in virtual worlds in the long term may approach or exceed the quality of life in nonvirtual worlds. Eventually, a virtual home may be as good or better than a nonvirtual home. In principle, virtual islands are on a par with nonvirtual islands, and virtual clothing is as effective as nonvirtual clothing. *As a result, virtual abundance has the potential to eliminate a great deal of distributive injustice.* Following David Hume, Rawls said that *scarcity is a condition for justice.* This means that without scarcity, the principles of justice do not apply. In conditions of abundance, there is no need for justice. The world might have other problems, but at least where considerations of distributive justice are concerned, a world with abundance has no flaws that need correcting. That raises the intriguing possibility that in the long term, virtual abundance could yield a sort of utopia, at least where distributive justice is concerned. Under virtual abundance, important material goods in virtual worlds are instantly reduplicable and available to all. This is a virtual version of what is sometimes called a post-scarcity society. (Chalmers, 2022, p. 362)

Sure enough, this might be entirely infeasible, not only due to technological constraints. Before, I have insinuated that human psychology seems to generate at least in some instances the desire for things that are exclusive and scarce, and more abundance is unlikely to dissolve those psychological mechanisms. However, this response again mistakenly measures the value of the narrative in terms of its feasibility, rather than in terms of its creative input to rethink the problem at hand. It is misguided to disregard how the vision could conduce to rethinking the entire problem of scarcity by proposing various technologies as a new nexus for a solution.

5.3 Reconciling Pluralities of the Good Life in Virtual Reality

Barbara Goodwin has associated the idea of utopianism with the act of conjuring images of the good life. She suggests that "the utopian [can be] defined as a social critic who argues from a vision of the Good Life, and tries to promote social change through the device of an alternative construction of society. Integral to this process is the implicit or explicit propagation of an alternative world-view" (Goodwin, 1980, p. 385). Much in her view has been endorsed in the previous argumentation: Constructions of alternative societies, as we have seen, can fulfill important cognitive and dispositional functions. Though, I have not alluded much to the concept of the good life. One could perhaps understand a posthuman being straightaway as a being who lives a good life. Is that entailed? The invocation of the good life brings us back to an earlier tenet: Given the current plurality of such visions of the good life, is not the hope to ever be able to reconcile them in one society, doomed to fail from the start? The tensions between different concepts of the good life are palpable within many contemporary societies: A devote Islamist might find inacceptable to be employed by a female CEO, a polygamist find inacceptable to live in a community of conservative puritans, a nudist finds his liberties undermined by laws that prohibit nudity in public. In many liberal societies, the coexistence of such diverse conceptions of the good life is only achieved by curtailing some peoples' rights to act out their convictions and preferences: The Islamist has no claim against a female boss and nudists and polygamists must defer their sexual activities into the realm of the private. Since the curtailment of some liberties seems to be the only way to achieve the reconciliation between these different conceptions of the good life, and since this seems *prima facie* undesirable and is factually undesired, what would a utopia look like in which reconciliation of different conceptions of the good life are achieved without curtailments of liberty? Robert Nozick has suggested that such a utopia is best understood as a meta-utopia, a world in which various communities coexist in a stable equilibrium. In each of these communities, different conceptions of the good life can be enacted. A stable state is reached when none of its inhabitants can conjure better communities than the ones already existing and between which they must be allowed to move freely back and forth[5]:

> The conclusion to draw is that there will not be one community existing and one kind of life led in Utopia. Utopia will consist of utopias, of many different and divergent communities in which people lead different kinds of lives under different institutions. Some kinds of communities will be more attractive to most than others; communities will wax and wane. People will leave some for others or spend their whole lives in one. Utopia is a framework for utopias, a place where people are at liberty to join together voluntarily to pursue and attempt to realize their own vision of the good life in the ideal community but where no one can *impose* his own utopian vision upon others. (Nozick, 1975, p. 311)

One could posit that virtual reality technologies enable realizing the central conditions of Nozick's account of utopianism: mobility between communities and the ease of concocting and creating new worlds with different makeups and rules. Virtual reality might be, therefore, appear as an experiment in meta-utopianism: The inhabitants can either chose freely to settle for any of the already existing communities or be given the power to create new ones. The emerging virtual world would consist of many different communities with varying socio-political and cultural texture and different norms and values. Twenty years ago, as if anticipating the emergence of the LGBTQ+ dating app *Grindr*, Larry Gross has suggested that the internet can be a place for community building for different sexual minorities (Gross, 2004). Aside from conflicting values that make impossible the simultaneous enactment of different ideals of the good life in reality, there are other constraints that could be overcome in the virtual world: Some people might have wanted to become a firefighter, but have latex allergy, or they might have wanted to become a body builder but are prone to injuries, etc. (Bainbridge, 2013). John Danaher reflects soberly that life in virtual reality will not be entirely harmonious, peaceful and

[5] The requirement is much stronger than it perhaps initially appears: It presumes that the good life can exist in a world in which other conceptions of the good life are enacted that might be detrimentally opposed without leading to any practical conflicts. The communities must leave each other be, while their members can fluctuate between them. And, naturally, Nozick's emphasizes the importance of liberty, which would require communities to discharge their members of any commitments toward them, allowing them to leave whenever they want. Together, these ideas (complete mobility and complete indifference to other communities) make a highly unrealistic pair.

without conflict. The mobility between different worlds might induce a sense of fleetingness and anonymity that even lowers the threshold for moral digression—as is common in internet forums today. He suggests that "virtual worlds are [no] playgrounds of immorality" and that "some actions in computer-simulated spaces have spillover effects in the real world" (Danaher, 2022, pp. 519 f.). However, as we have seen before, he and Chalmers remain optimistic: "In principle, VR can be much more than escapism. It can be a full-blooded environment for living a genuine life" (Chalmers, 2022, p. xvii).

References

Bainbridge, W. S. (2013). Transavatars. In M. More & V.-M. Natasha (Eds.), *The Transhumanist Reader* (pp. 91–99). Wiley.

Bostrom, N. (2003). Human Genetic Enhancements: A Transhumanist Perspective. *The Journal of Value Inquiry, 37*(4), 493–506. https://doi.org/10.1023/B:INQU.0000019037.67783.d5

Castle, N. (2022). In Vitro Meat and Science Fiction. *Extrapolation, 63*(2), 149–179. https://doi.org/10.3828/extr.2022.11

Chalmers, D. J. (2022). *Reality+: Virtual Worlds and the Problems of Philosophy* (1st ed.). W. W. Norton.

Danaher, J. (2022). Virtual Reality and the Meaning of Life. In I. Landau (Ed.), *The Oxford Handbook of Meaning in Life* (pp. 508–524). Oxford University Press.

Frank, R. H. (2011). *The Darwin Economy*. Princeton University Press.

Goodwin, B. (1980). Utopia defended against the liberals. *Political Studies, 28*(3), 384–400. https://doi.org/10.1111/j.1467-9248.1980.tb00476.x

Gross, L. (2004). Somewhere There's a Place for Us: Sexual Minorities and the Internet. In M. Sturken, T. Douglas, & S. J. Ball-Rokeach (Eds.), *Technological Visions: The Hopes and Fears That Shape New Technologies* (pp. 255–269). Temple University Press.

Hauskeller, M. (2012). Reinventing Cockaigne: Utopian Themes in Transhumanist Thought. *Hastings Center Report, 42*(2), 39–47. https://doi.org/10.1002/hast.18

Hughes, J. (2004). *Citizen Cyborg: Why Democratic Societies Must Respond to the Redesigned Human of the Future*. Westview Press.

Huxley, J., & Birx, H. J. (Eds.). (1992). *Evolutionary Humanism*. Prometheus Books.

Kallis, G., & March, H. (2015). Imaginaries of Hope: The Utopianism of Degrowth. *Annals of the Association of American Geographers, 105*(2), 360–368. http://www.jstor.org/stable/24537850

Kitcher, P. (1996). *The Lives to Come: The Genetic Revolution and Human Possibilities*. Allen Lane.
Lessig, L. (1999). *Code and Other Laws of Cyberspace*. Basic Books.
Nagel, T. (1991). *Equality and Partiality*. Oxford University Press.
Nozick, R. (1975). *Anarchy, State, and Utopia*. Blackwell.
Parrinder, P. (1997). Eugenics and Utopia: Sexual Selection from Galton to Morris. *Utopian Studies, 8*(2), 1–12. http://www.jstor.org/stable/20719681
Pleij, H., & Webb, D. (2001). *Dreaming of Cockaigne—Medieval Fantasies of the Perfect Life*. Columbia University Press. https://doi.org/10.7312/plei11702
Rutherford, A. (2023). *Control: The Dark History and Troubling Present of Eugenics*. W. W. Norton.
Sen, A. K. (2010). *The Idea of Justice*. Penguin Books.
Shew, A. (2020). Ableism, Technoableism, and Future AI. *IEEE Technology and Society Magazine, 39*, 40–85. https://doi.org/10.1109/MTS.2020.2967492
Smith, S. E. (2021, February 14). Disabling Utopia to Save It. Society Can Benefit from Conjuring Worlds That Model Diversity and Inclusion. *The Nation*. https://www.thenation.com/article/society/disability-utopia/
van Wynsberghe, A. (2021). Sustainable AI: AI for Sustainability and the Sustainability of AI. *AI and Ethics, 1*(3), 213–218. https://doi.org/10.1007/s43681-021-00043-6
Weindling, P. (2012). Julian Huxley and the Continuity of Eugenics in Twentieth-Century Britain. *Journal of Modern European History, 10*(4), 480–499. https://doi.org/10.17104/1611-8944_2012_4_480

CHAPTER 6

Conclusions

Abstract The final chapter synthesizes the main threats of the previous and recaps the main conclusions. Since my defense of technological utopianisms is conditional to their ability of advancing our understanding of justice, one might wonder—if justice is so broadly construed—whether any fail to do so. I will also suggest that the methodological point developed here requires us to accept the existence of evil utopias, a fate it shares to the same degree with more realistic approaches. The need and desire for utopianism will always embark on a particular perception of the unacceptability of the present.

Keywords Engagement with bad utopias · Space tourism · Nazi utopias · Prospects for future research · Utopia and the present

The examples introduced in the previous chapter underscore the idea that technological utopias advance our understanding of justice by innovating and challenging convictions about its scope and location. The list of examples is meant to be instructive and is far from complete. One might add cyber-utopianism—a view that sees the emergence of virtual communities as a stimulant for more empathy toward our distant others—as another example of this general idea. Various authors have pointed out that our lack of support for distant others across the globe (and in the future) might be a product of our evolution from small communities, in which

M. Sand, *Technological Utopianism and the Idea of Justice*,
https://doi.org/10.1007/978-3-031-75945-1_6

actions had limited causal consequences (Persson & Savulescu, 2012). While the latter has changed, the former has not. Promoters of moral enhancements like Persson and Savulescu fancy locating a solution to this problem again in our bodies and suggest intervening with pills and drugs to increase empathy, overcome status *quo bias* and make us more disposed to aid. The "internet imaginaire", on the other hand, proposes a solution elsewhere: It fancies the potential for overcoming partiality and partisanship through wider access to information about distant others and the possibility of communicating with them, which should lead to increasing feelings of closeness and connectedness on a global scale—thence the idea of the global village (Daub, 2020). As Frances Cairncross suggests: "Bonded together by the invisible strands of global communication, humanity may find that peace and prosperity are fostered by the death of distance" (quoted in: Flichy, 2007, p. 87).

In reality, virtual communities might not be a silver bullet to eliminate partiality, partisanship and global inequality. We might end up—as usual—by largely accepting the *status quo* of global inequality and muddling through in small steps combining a plethora of different approaches; continuing to appeal to individual duties and responsibilities, utilizing education, tools of advertisement or political institutions (tax reductions for donations), next to investments into various technologies. But that does not undermine the cognitive and dispositional power that technological utopias such as those discussed in the present essay harbor. With some of those narratives, we have become so acquainted that we have forgotten that they are the product of a laudable, creative act of the imagination and that that means they have in some sense already changed the world around us.

I shall conclude this essay with two final remarks—an open question and a possible objection to my approach. If it is true that my arguments are modest in that they support study and engagement with utopianism if those advance our scope of justice and given the broad ideas of the scope and the location of justice that I have presented in Chapters 4 and 5, I might owe the reader an example of a technological utopia that is devoid of justice reflections. This is not an easy task indeed. We have seen that technological utopias exceed the artificial dichotomy between individualist and collectivist utopias and between the social and the political. I would be inclined to consider the *Astrotopias* of colonizing Mars and space tourism as narratives that are almost completely devoid of justice reflections. Those have been named utopias, too (Rubenstein,

2022; Tutton, 2021), but they are not concerned with how people live better together. Mary-Jane Rubenstein points out the hollow motives that underlie those visions:

> Rather than proposing an alternative to the extraction of 'resources,' the relentless pursuit of profit, and the wasteful cruelty of factory farming; rather than using his prodigious intellect to solve the problem of food distribution or his prodigious fortune to seed a universal basic income (or to pay a few dollars in federal taxes), Bezos is spending his money and time exporting the whole damned system into space. The alternative would be 'stasis,' or even reversal. And Bezos wants to keep moving 'forward,' so he's going to have to go up and away. (2022, p. 20)

Rubenstein's judgment is piercing and understandable. The utopia of space tourism, in particular, seems extravagant and devoid of justice considerations. Still, one might say that space colonization, in contrast, underscores the importance of the continuation of human existence as a *sine qua non* for a just society. For any just society to exist, it is necessary that humans survive climate change and other existential threats. Perhaps that is best achieved by colonizing other planets. Due to the empirical nature of the latter assumption, I cannot judge whether that is a warranted train of thought. The general difficulty of finding utopias that are devoid of justice considerations merely underscores the incompleteness of our understanding of this normative idea, which underlies the central claim of this book: Utopianism conjures infinite ways of living and being better in the future. It would be aloof to reject their creative potential in advancing and expanding our imperfect view of a just society, how it might be brought about and how it might be shaped. At the fore of the present project was the attempt to tie insights from Utopian Studies and philosophy of technology closer to ongoing debates in political philosophy by underscoring that technological utopias scrutinize the location and scope of justice and, thereby, exceed the narrow focus on justifications of principles of justice. For a more comprehensive, future account one could characterize this as the ability to re-envision and innovate values more broadly, in the above example the disvalue of extinction, for instance.

It is important to underscore—at pain of repeating myself—that my defense provides a reason for engagement with some technological utopias insofar as they have the capacity to advance our idea of the scope and location of justice. It is *not a reason* for acceptance or promotion of any of those visions, let alone for attempting to realize them. Utopias must be part of the conversation, where we want to head toward. They should not be rejected for being unrealistic or for being flawed in other ways that have been discussed before, e.g. for being perfectionist. Even if they were unrealistic or perfectionist, they might still provide valuable insights in which ways the present is flawed and needs change and perhaps provide some ideas on how to bring this change about.

This brings me to my second point—another possible objection to my overarching approach. Defending utopianism in the way might motivate the following response: One might say that in their propensity to advance our idea of justice, its scope and location, even abhorrent and evil utopian visions might be worthwhile of engagement. At the very least, I would have to welcome their existence. Consider, Nazi utopias of living in the prairies of the East or in the North (Stratigakos, 2022). The Nazis joggled grandeur visions of living in the vast prairies of the East, weirdly combining in these fancies of high-tech militarization with pagan mythology—blood and ground—to connect farming and peaceful, ethnically and socially cleansed Arian village life (Fest, 1991, pp. 52 f.). These narratives have been considered as utopias, too, perhaps with good reasons. Must I consider it somehow as good that these visions are around, and are they also worthy of engagement, thus? Given my lack of commitment to a definition of utopia, I cannot deny on grounds of definition the applicability of the utopian concept to those visions, although my bowel revolts when thinking of such images of ethnically cleansed communities as a better way of living and being. Still, I will bite the bullet and agree that it is somehow good that even such abhorrent and evil visions are around. However, it is crucial to be clear what my modest vindication of technological utopianism entails and what not. I have not defended utopianism *tout court*. I have defended utopianism and in particular technological utopianism only insofar as they advance our idea of the scope and location of justice. Nor have I defended any particular utopias' view on justice. It is not implied in my view that any particular (technological) utopia proposes the right ways of thinking about justice, let alone presenting a demand to be realized.

If a defense of utopianism would imply a commitment to some particular utopias, including bad ones, it would be an implication of realistic methods and outlooks on socio-political justice, too. Realism, as a broad stance on what kind of political ideas are suitable to base political decisions on and design socio-political institutions, is equally a defense of all sorts of diverging realistic proposals: The very realistic proposals of ridding immigrants of their political rights, closing our borders and starting wars with other nuclear powers, are morally no less alarming than some abhorrent utopian visions. Without embarrassment, realists might say that such proposal retain an advantage over utopian proposals for political theorizing, without suggesting that that they represent the right way of doing politics or contain the right ideas of justice. In short, the problem of evil political ideas and bad political proposals runs diagonal to the discussion what methodology to employ when thinking about justice—a realistic or an idealistic one: Either of these perspectives refers to a methodological commitment, not a substantial one. If the objection is leveled against one of these, it will have to be leveled against the other, too.

As I have underscored various times throughout this essay: The optimism that technological utopias place in the power of technology might be considered just as naïve, as the belief in the power of socio-political institutions or the benevolence of individuals to create a more just society. A cursory look into human history provides plenty of reasons to doubt that radical progress is feasible by way of human effort, morality and politics. Hence, rejecting technological utopianism for naiveté drags utopianism in general down the drain with it. The only viable alternative seems to be to settle for a more modest realism. However, I am not convinced that realism is necessarily a more comfortable, less intrusive or less violent way of living. The idea that someone's utopia is another person's dystopia is catchy, yet one-sided. The notion underplays the silent violence of the *status quo* of current societies. One might rhetorically contrapose: Someone's liberal democracy is someone else's inegalitarian, white patriarchy. Chrostowska begins her analysis pondering the question "Is this dystopia?"—which is sprayed on a wall that she passes by (Chrostowska, 2021, p. 1). If we turn toward utopia out of frustration about the felt inadequacies and unacceptability of the present, we do so because of an unwillingness to settle with a society "[that] may fail to embody any ideal at all", as Nagel puts it (Nagel, 1989, p. 21). Discussions of utopia continue to expose how we diverge in our perceptions of the present and its shortcomings and in the conclusions, we draw from those diverging perceptions. Utopianism is as much part of an endeavor to understand the present as it is an endeavor to guide us towards a better future.

References

Chrostowska, S. D. (2021). *Utopia in the Age of Survival—Between Myth and Politics*. Stanford University Press.

Daub, A. (2020). *What Tech Calls Thinking—An Inquiry into the Intellectual Bedrock of Silicon Valley*. Farrar.

Fest, J. (1991). *Der zerstörte Traum - vom Ende des utopischen Zeitalters*. Siedler.

Flichy, P. (2007). *The Internet Imaginaire*. MIT Press.

Nagel, T. (1989). What Makes a Political Theory Utopian? *Social Research, 56*(4), 903–920. www.jstor.org/stable/40970571

Persson, I., & Savulescu, J. (2012). *Unfit for the Future: The Need for Moral Enhancement* (1st ed.). Oxford University Press.

Rubenstein, M.-J. (2022). *Astrotopia: The Dangerous Religion of the Corporate Space Race*. The University of Chicago Press.

Stratigakos, D. (2022). *Hitler's Northern Utopia: Building the New Order in Occupied Norway*. Princeton University Press.

Tutton, R. (2021). Sociotechnical Imaginaries and Techno-Optimism: Examining Outer Space Utopias of Silicon Valley. *Science as Culture, 30*(3), 416–439. https://doi.org/10.1080/09505431.2020.1841151

Index

A
agent of change, 100, 135
approximation fallacy, 44
artificial intelligence, 17, 18, 54, 76, 85, 119

B
background assumptions, 1, 2, 46, 47, 60, 61, 110, 111, 122–125, 127
basic structure, the, 92, 111–113, 117, 119, 120, 139
blueprint, 11, 30–32, 35, 41, 87, 121
blueprinting, 32
boredom, 30, 53, 54, 56, 59, 123
Bostrom, Nick, 17–19, 22, 23, 53, 75, 76, 78, 95, 96, 99, 134–136

C
capitalism, 23, 24, 51, 52, 73, 86, 90, 93, 95–99, 111, 121, 140
Chalmers, David, 20, 21, 77, 141, 144
Chrostowska, S.D., 51, 52, 72, 88, 151
climate change, 38, 40, 49, 51, 87, 89–91, 149
coercion, 34, 37, 42, 56
Cohen, G.A., 46, 55, 113, 114, 119–121, 123, 124
cyber-utopianism, 22, 91, 147

D
Danaher, John, 19, 20, 53–57, 69, 77, 83, 123, 143, 144
democracy, 22, 24, 25, 35, 77, 78, 99, 101, 151
dystopia, 24, 56, 151

E
elitism, 52, 72
enhancement, 17–19, 79, 84, 99, 101, 134, 135, 148
epistemological absolutism, 59, 64
equality, 50, 96, 98, 99, 120, 132
Estlund, David, 12, 42–46, 64, 110
eugenics, 72, 85, 135, 137

M. Sand, *Technological Utopianism and the Idea of Justice*,
https://doi.org/10.1007/978-3-031-75945-1

F
functions of utopia, 47
  cognitive function, 2, 4, 42, 89, 110
  critique, 51, 53, 109
  dispositional function, 4
  estrangement, 49, 99
  subversive function, 43, 47, 51, 86, 97, 99, 110

G
genetics, 17, 133–135
the good life, 2, 31, 62, 142, 143
Goodwin, Barbara, 37, 39, 41, 62, 142

H
historicism, 33
Hughes, James, 80, 96, 97, 99–101, 124, 133, 135
human nature, 1, 2, 17, 18, 46, 47, 82, 84, 85, 93, 101
Huxley, Julian, 16, 18, 56, 73, 136

I
idealism, 38, 39, 43, 46, 62, 63, 74, 85, 98, 151
  unrealistic, 46, 63, 64
ideologically imprisonment, 43, 76, 102, 109, 135
immortality, 72, 81–83, 87, 88, 124, 125
individual behavior, 3, 4, 52, 82, 92, 93, 102, 111, 113–117, 119–121, 127

J
Jameson, Frederic, 5, 49, 64, 72, 109
justice
  location of justice, 1, 13–15, 46, 92, 95, 111, 120, 122, 125, 131, 133, 139, 148–150
  principles of justice, 13, 43, 44, 46, 62, 64, 110, 112, 141
  scope of justice, 3, 47, 52, 60, 61, 70, 117, 121–123, 139, 148, 149
  theories of justice, 1, 2, 42–47, 52, 59, 61, 64, 92, 111, 123–127, 132

K
Kumar, Krishan, 5, 6, 12, 41, 42, 47, 74
Kurzweil, Ray, 33, 78

L
Land of Cockayne, 12, 80, 138, 139
Levitas, Ruth, 4, 7–10, 12, 17, 50, 54, 80, 102, 109
Levy, Jacob, 2, 123–127
liberal capitalism, 51, 93, 95, 98
liberal democracies, 37, 38, 99
libertarianism, 22, 39, 99, 138, 139
limitarianism, 139
limited altruism, 2, 101, 111, 124

M
Mao, Douglas, 40, 101, 102, 121
moral enhancement, 101, 148
More, Max, 16, 31, 33, 59, 75, 76, 80, 99
More, Thomas, 5, 6, 23
Morozov, Evgeny, 70, 89–91
mortality, 2, 122–125
Mosco, Vincent, 21, 22, 71–73
motivational constrains, 47

N
Nagel, Thomas, 47, 132, 133, 151
naiveté, 87, 88, 90, 91, 151
Nordmann, Alfred, 53, 79–84
Nozick, Robert, 142, 143

O
object of change, 52, 101, 135

P
perfectionism, 13–15, 17, 29–32, 53, 55, 59, 61
philosophy of technology, 4, 15, 85, 149
piecemeal engineering, 11, 34–39
pluralism, 14, 59, 111, 138
political philosophy, 4, 15, 17, 42, 43, 61, 85, 109, 127, 138, 149
Popper, Karl, 11, 29–42, 52, 53, 62, 69, 83, 124
postphenomenology, 115
practical relevance, 42, 43, 47, 52, 88, 126
prediction, 23, 31, 33, 73, 79–81, 83

R
Rawls, John, 43, 46, 61, 111–114, 117–121, 127, 132, 139, 141
realism, 74, 89, 102, 119, 141, 150, 151
reasonable disagreement, 2, 14, 61, 62

S
Saage, Richard, 16, 41, 51, 52, 75, 80, 83, 90, 94, 95, 97–101, 110, 111, 121, 135
Sargisson, Lucy, 5–7, 13, 23, 31, 42, 49, 59, 75, 78, 109
scarcity, 2, 14, 60, 61, 111, 125, 137, 138, 141
Scheffler, Samuel, 82, 112–114, 119
Segal, Howard, 6, 7, 11, 15, 72–74, 77, 88, 100
Sen, Amartya, 59–64, 69, 137, 138
socialism, 120, 121
social obliviousness, 3, 4, 86, 89, 93, 111
socio-political institutions, 4, 14, 52, 55, 56, 59, 92, 93, 100, 102, 111–115, 118–121, 127, 133, 135, 140, 150, 151
socio-technical systems, 119
stagnation, 15, 30, 54–57, 59, 123
sufficientarianism, 139

T
Thaler, Mathias, 4, 40, 49–51, 88, 109
totalitarianism, 11
Transhumanism, 16–18, 22, 31, 33, 51, 53, 59, 72, 75, 76, 78, 85, 87, 93–99, 111, 133, 135, 137–139

U
utopia
- collectivist, 3, 101, 121, 148
- individualist, 3, 101, 121, 148
- intrinsic value of, 43
- practical relevance, of, 42, 43, 52
- traditional, 15, 41, 43, 51, 54, 72–74, 97, 100–102, 121

utopianism
- anti-anti-utopianism, 109
- anti-utopianism, 4, 17, 29, 30, 32, 53, 64, 69, 73, 89, 109
- technological utopianism, 1–4, 6, 9, 10, 14, 29, 30, 44, 47, 52, 53,

64, 69, 70, 73, 86, 88, 93, 122, 131, 150, 151
Utopian Studies, 4, 8, 15, 43, 70, 85, 109, 149

V
violence, 14, 15, 30, 34, 37, 40–42, 52, 76, 77, 151
virtual reality, 19–21, 50, 62, 77, 87, 141–144

W
Winner, Langdon, 22, 72, 86, 89

The manufacturer's authorised representative in the EU is Springer Nature Customer Service Centre GmbH, Europaplatz 3, 69115 Heidelberg, Germany. If you have any concerns regarding our products, please contact ProductSafety@springernature.com

Printed and bound by CPI Group (UK) Ltd, Croydon, CR0 4YY
01/10/2026
02225978-0012